P9-CJV-601

Biomedicalization

of

Alcohol
Studies

Biomedicalization

of

Alcohol Studies

Ideological Shifts and Institutional Challenges

Lorraine T. Midanik

ALDINETRANSACTION
A Division of Transaction Publishers
New Brunswick (U.S.A.) and London (U.K.)

Copyright © 2006 by Transaction Publishers, New Brunswick, New Jersey.

All rights reserved under International and Pan-American Copyright Conventions. No part of this book may be reproduced or transmitted in any form or by any means, electronic or mechanical, including photocopy, recording, or any information storage and retrieval system, without prior permission in writing from the publisher. All inquiries should be addressed to AldineTransaction, A Division of Transaction Publishers, Rutgers—The State University, 35 Berrue Circle, Piscataway, New Jersey 08854-8042. www.transactionpub.com

This book is printed on acid-free paper that meets the American National Standard for Permanence of Paper for Printed Library Materials.

Library of Congress Catalog Number: 2005057077
ISBN: 0-202-30835-9
Printed in the United States of America

Library of Congress Cataloging-in-Publication Data

Midanik, Lorraine T.
    Biomedicalization of alcohol studies : ideological shifts and institutional challenges / Lorraine T. Midanik.
        p. cm.
    Includes bibliographical references and index.
    ISBN 0-202-30835-9 (alk. paper)
    1. Alcoholism—Research—Philosophy. 2. Alcoholism—Physiological aspects. 3. Alcoholism—Genetic aspects. 4. Alcoholism—Social aspects. 5. Neurobiology—Philosophy. I. Title.

RC565.M432   2005
362.292'01—dc22   2005057077

This book is dedicated to two extraordinary women who passed away in March 2005: Sheryl Kramer and Joan Dunkel. Both of them graced the world and made it a better place. I loved them, learned from them, and miss them daily.

Sheryl Kramer
1946-2005

Joan Dunkel
1928-2005

Contents

Acknowledgments

This book has taken a long time to complete. Over the years it has shifted from a book on self-report validity measures, to a book about how and why social science researchers "symbolically" establish validity to look as if they are "real" scientists, to a book on the general trend of biomedicalization and how the alcohol field, as a case study, has been affected by it. As one might imagine, there have been many people along the way who have helped me shape and mold this book into what it finally is today.

Probably the strongest influence in my career on how I view and examine alcohol issues has come from the researchers who were associated with the Social Research Group (renamed the Alcohol Research Group (ARG) in 1981) when I arrived as a post-doctoral fellow in the School of Public Health at the University of California at Berkeley in 1979. I was fortunate to obtain excellent training and support from its Director, Robin Room, as well as from Walter Clark and Kaye Fillmore. Walt Clark was then and still is my mentor who helped shape my thinking about how one defines alcohol problems as well as the importance of "drunkenness" – as a measure, of course! Andrea Mitchell was kind enough in 1979 to share her office with me in the library and has, over the years, helped me in so many ways to obtain the materials I have needed for this book. I am still affiliated with ARG and want to thank Tom Greenfield, its current Director, for his support over the years and for our multiple joint efforts that primarily focus on improving methods in alcohol research.

At the School of Social Welfare at the University of California at Berkeley, I have also received guidance and support particularly from Dean James Midgley whose sage advice and gentle push for me to finish this book was very much appreciated. I was also very much encouraged by Professors Eileen Gambrill and Mac Runyan who are examining the same issue in different fields. I also want to thank Dina Redman, Melissa Martin-Mollard, and Christine Lou for their assistance on this book, and Sharon Ikami for her secretarial help. I also want to acknowledge the patience of the many students who, for many years, heard about biomedicalization in my health policy and drug and alcohol policy classes.

In Spring 2004, I was very fortunate to spend the semester in Stockholm at the Centre for Social Research on Alcohol and Drugs (SoRAD) at Stockholm University as a Fulbright Scholar where Robin Room was the Director. It was wonderful to have an opportunity to work with Robin again in a research environment that was both supportive and intellectually stimulating. During my time at SoRAD I conducted a research project that is included in Chapter 6, and I finished writing this book. I want to thank the entire staff at SoRAD for their feedback on my work. I particularly want to thank Robin Room, Börje Olsson, Fabian Sjö and Christel Hopkins for their help with my project and for making me feel so welcome at SoRAD. Finally, I would like to thank Jeannette Lindström and Maria Dahlrens of the Fulbright Commission in Stockholm whose support made this book possible.

Originally this book was to be published at Aldine de Gruyter under the guidance of Richard Koffler, the editor at that time. Richard's early advice in shaping this book was invaluable. In 2004, Aldine moved to Transaction Publishers, and I began working with Irving Horowitz, Mary Curtis, and Larry Mintz. They were wonderful and helped make this book stronger. I very much thank them for their help.

Finally, I have been blessed in my personal life to have terrific friends who enrich my life. I particularly want to thank Helen Shrager, Stephanie Coleman, Betsy Silver, Barbara Honig, the Gordon-Kirsch family, Carol and Carl Ingram, and Shelly Halperin for putting up with me for so many years; Mary Ciddio, Angela Quarteroli and Susan Brahan, my walking women's group; and, my Thursday night bridge group (you know who you are) who always make me laugh even when I make mistakes (I still know that deals, who shuffles, and who cuts!). Last, but clearly not least, my family has always been a caring source of inspiration for me; they truly believed I could finish this book. I want to thank Celia Blum, Larry, Corey and Jesse Kahanofsky and the Zauss family who are so important to me; my mother, Sadie Midanik, whose love and joy of life I cherish; and, my wonderful daughters, Beth and Amy, who make me realize everyday what is very important in life – they are truly my role models. Finally, I owe much more than thanks to my husband, Stephen Blum, whose encouragement and love make my world complete. How lucky I am to have him as my partner. Truly if it were not for Stephen, the book would not have been written. This book honors him.

Permissions

Material in chapter 7 is reprinted from *Social Science and Medicine,* volume 60, L. Midanik and R. Room, "Contributions of social science to the alcohol field in an era of biomedicalization," pp. 1107-1116, 2005 with permission from Elsevier.

Material in chapters 4 and 5 is reprinted from the *Journal of Public Health Policy,* volume 25, L. Midanik, "Biomedicalization and alcohol studies: Implications for policy," pp 211-228, 2004. Reprinted with permission.

Figure 3.1 is reprinted from *The British Journal of Addiction,* volume 54, pp. 133-148, 1958 (presently *Addiction*) with permission from Blackwell Publishers, Ltd.

The Twelve Steps and Twelve Traditions (Tables 3.1 and 3.2) are reprinted with permission of Alcoholics Anonymous World Services, Inc. ("A.A.W.S.") Permission to reprint the Twelve Steps and Twelve Traditions does not mean that A.A.W.S. has reviewed or approved the contents of this publication, or that A.A. necessarily agrees with the views expressed herein. A.A. is a program of recovery from alcoholism only—use of the Twelve Steps and Twelve Traditions in connection with programs and activities which are patterned after A.A., but which address other problems, or in any other non-A.A. context, does not imply otherwise.

Table 3.3 is reprinted with permission from the *Quarterly Journal of Studies on Alcohol,* volume 7, pp. 1-888, 1946 (presently *Journal of Studies on Alcohol*). Copyright *Journal of Studies on Alcohol,* Inc. Rutgers Center of Alcohol Studies, Piscataway, NJ 08854.

1

Introduction:
The Era of Biomedicalization

"There's a lot of money to be made from telling healthy people they're sick. Some forms of medicalising ordinary life may now be better described as disease mongering: widening the boundaries of treatable illness in order to expand markets for those who sell and deliver treatments. Pharmaceutical companies are actively involved in sponsoring the definition of diseases and promoting them to both prescribers and consumers. The social construction of illness is being replaced by the corporate construction of disease." (Moynihan, Heath & Henry, 2002, p. 886)

What is Biomedicalization?

In June 2000, I was invited to a small meeting of social work faculty who teach in the area of alcohol and drugs. The meeting, entitled "Social Work and the Neurobiology of Addictions," was funded jointly by the Texas Commission on Alcohol and Drug Abuse and the Waggoner Center for Alcohol and Addiction Research at the University of Texas at Austin. I felt some trepidation attending this meeting given that I know little about neurobiology; however, I was intrigued by the opportunity to learn more about the possible applications of this science to the social work profession. Eminent scholars in neurobiology gave complicated yet understandable lectures on the "science" of addiction; other speakers spoke of the need to integrate neurobiology in the courses that we routinely teach on substance abuse methods and policy within our academic institutions.

While the meeting provided some impetus for thinking about how neurobiology as a method and as a discipline can be incorporated into our current classes and research, I was struck by the recurring theme and belief that "addiction is a brain disease." This point of view was presented not as a question for inquiry, but as a self-evident, universal truth not to be questioned. It was stated with virtual certainty that addiction, defined as DSM-IV Dependence (APA, 1994),

1

was not a matter of morality, or, for that matter, mental illness, but rather could be more accurately understood as a brain disease for which neurobiology could best understand and treat.

As someone who has been in the alcohol and drug field for many years, I was somewhat surprised by the fervor of this assertion, and I was also saddened by what I thought the consequences of an uncritical acceptance of this brain disease model would be for research, treatment, and policy in the alcohol field. That alcohol and drug addiction is presented in the framework of disease is certainly not a new idea; yet, this powerful and rapid shift towards defining it *solely* in terms of brain function or dysfunction is new and has quite significant ramifications for research, treatment, and prevention. I sat at the meeting wondering what happened to our broader understanding of addiction when such reductionistic statements can be made without being carefully examined and debated.

Declaring addiction a "brain disease" carries with it several advantages for specific vested interest groups. These include a stronger focus on neurobiological and genetic mechanisms for diagnosing, treating, and possibly preventing alcohol and drug problems, thus setting the stage for examining and treating these issues from an almost always highly individualistic perspective. Yet, the presumption that addiction issues are wholly, or even mostly, defined by brain dysfunction alone is alarming in that it severely limits how alcohol and drug problems are viewed and correspondingly, which set of disciplines contribute most to the field. By defining addiction as a "brain disease" and hence individualizing the issue, the importance of *context and the environment* that indeed constantly and dynamically interact with the brain are essentially ignored. To state emphatically that addiction is largely a "brain disease" is as narrow and unproductive as stating that addiction is largely a social or environmental phenomenon. Both the physical and the environment are important dimensions, and to ignore either and the relationship between the two can cause more problems than it simplistically purports to "solve." Moreover, the emphasis on addiction minimizes the vast majority of alcohol and drug-related problems that are *not* entirely the result of addiction per se, for example, alcohol-related traffic crashes and underage drinking. These pervasive problems have a pronounced impact on our society and in other countries and societies worldwide.

Around the same time that this meeting took place media sources such as newspapers, professional journals, and television news shows were beginning to report on the preliminary results of the Human Genome Project (HGP) that had been initiated in 1986. The main goal of the HGP is to map fully the human genetic structure so that, it is hoped, prevention and even eradication of the over 4,000 genetic diseases may occur (Conrad & Gabe 1999). With this promise of finding solutions to major disease entities, the HGP and genetic research in general has enjoyed tremendous public and governmental support. Yet, this arena, referred to as "The New Genetics," is not without its own significant ethical issues, dilemmas, and needs for careful assessment in a highly important set of domains including the politics of public versus private genetic research, threats to privacy, and issues of potential discrimination particularly with regard to obtaining employment and health/life insurance (Cunningham-Burley & Kerr, 1999). Perhaps one of the most disturbing effects is an increased and more intense focus on the genetic and biological basis for disease by the media, and its spillover into areas that have both social and genetic components such as schizophrenia, manic depression and alcoholism (Allen, 1999). This expansion of the explanatory domain for genetics appears to result in the consequent exclusion of environmental and social concerns, a process that is sometimes referred to as "genetic determinism" or "geneticism." Conrad (1999) has elaborated further on this trend by using the acronym "O-GOD" (one gene, one disease) to illustrate the reductionistic and thus simplistic line of thinking that "...a direct and virtual one-to-one relationship [exists] between genetics and behaviour" (p 3). Regardless of the terminology, the entire social, political, and scientific process of the New Genetics is accompanied by very high expectation, worldwide attention, and fiscal support.

While the meeting in Texas in 2000 and the multitude of recent reports on the progress of HGP made me increasingly aware of the growing emphasis of the biomedical world in the health and social sciences, it can be argued that these are indicators of a much larger trend that has been occurring since the mid-1980s, termed by some researchers as biomedicalization. This change is characterized by a new wave of individual level explanations for behaviors. Unlike the concept "medicalization" that referred to the growth of medicine's authority over traditionally non-medical areas of life such as juvenile delinquency and sexual orientation (Conrad & Schneider, 1980), biomedicalization offers more complexity in that it encompasses new "...social forms and practices of a highly and increasingly technoscientific biomedicine" (Clarke, Shim, Mamo, Fosket & Fishman, 2003). This process of biomedicalization reduces the unit of analysis of research, treatment, and policy to biomedical conditions or processes, often genetic, within which an individual is the locus of specific "problems." Biomedicalization assumes that an individual's genetic and physical being explains virtually the entire etiology of the "disease" and hence dictates its course of treatment. The development and implications of medicalization and biomedicalization are discussed in more detail in chapter 2.

Why is Biomedicalization an Important Concern?

In and of itself, biomedicalization is not inherently a bad or a good process. It can clearly be seen in the context of scientific progress—the next step in advancing a better means to diagnose, treat, and prevent disease. To the extent that it is a major social trend, biomedicalization is evident in a variety of fields such as psychiatry, medicine, aging and women's health (discussed in chapter 2) and can be characterized by the increased reliance on new technology and medications, the need to individualize social problems, and the shift away from social factors as significant components of our understanding of health or disease. As suggested by the quote at the beginning of this chapter, there are powerful financial incentives to support illness and disease as individual level problems that can be remedied by individual, pharmacological solutions. Glenn (1988) further points to factors such as the economic structure of American medical practice, the reimbursement process for American health care providers who have come to rely largely on biologically based codes to the near exclusion of any other factors, as well as the larger population's definitions of health and disease to explain why the biomedicalization process has so much support in the U.S.

The alcohol field, and particularly how alcoholics have been defined and treated, is an excellent case example of both medicalization and biomedicalization. As will be discussed in more detail in chapter 3, alcohol issues have historically undergone major shifts ranging from being seen as moral and legal problems to medical diseases (Siegler, Osmond & Newell, 1968). Currently, the field is influenced by the increasing dominance of the biomedical model which promises not only to be better able to detect this disease, but also to treat and potentially to prevent it (Midanik, 2004).

While it is not my contention that the influence of the biomedical model in the alcohol field and in other fields is inherently problematic when used in conjunction with social models, the premise of this book is that biomedical research has taken such a dominant role in several fields that it has marginalized other methods or perspectives. This is highly problematic for several reasons.

First, while there is occasional mention of the importance of environment in this literature, all too often the bulk of biomedical research is focused on individuals—fairly (or sometimes even totally) independent of the environment in which they live. Thus, attention is placed on the brain, neurons, or genes, for example, and little, if any, focus is on the interactions of that brain, those neurons, or those genes within the physical, social, and political environment (Conrad, 1999). Second, biomedical research is predominantly concerned with fairly extreme forms of behavior, for example, alcoholics; the harm that is caused by the large numbers of other types of drinkers who are not alcoholics is minimized. Thus, biomedicalization almost completely ignores the importance of preventive efforts that are not focused on individuals per se, such as programs on the community or state levels. Consequently, the biomedicalization process becomes one which is more fully in line with market interests, such as the beverage industry in the U.S., that historically has not been supportive of social policy reforms that affect the population at-large such as increasing taxes or limiting access and availability (Morgan, 1988). Third, the overwhelming dominance of the biomedical movement in the alcohol field and in other fields has placed the social science disciplines working within this field in a counter-productive defensive position by imposing often quite limited and reductionistic biomedical standards (considered by some to be "higher standards") on all researchers. Under these dominant assumptions, it is generally expected that social science researchers adhere to these "more rigid" standards to obtain research grants, publish articles, etc. The use of biomedical standards in alcohol studies has several dimensions that impact social science research. First, they involve the general distrust of self-reports as a means of obtaining data and suggest that data must be obtained from so-called "objective measures," (e.g., breath tests, liver function tests) as outcome measures or as a means to determine the validity of the data that are being used. It has yet to be determined how much these "objective" measures contribute to our understanding of how respondents answer questions or to our ability to make a diagnosis (Babor, 2000). However, use of biomedical measures is symbolic of science per se and is thus often mandated in social science research in the alcohol field. Second, the promotion of clinical trials as the benchmark of research design minimizes or even eliminates as "unnecessary" the importance of other types of designs that can be both useful and important in advancing our understanding of harmful alcohol use and its subsequent problems. Finally, biomedical standards are generally dismissive of qualitative approaches to understanding alcohol use among a range of drinkers. Insistence on "precision" makes no allowance for other strategies that are often extremely important ways of getting particular information that cannot be obtained through other means. Of course, biomedical research methods are valuable; however, I am arguing that there is something inherently problematic about their current dominance in the alcohol field because of the resultant marginalization of other, equally important perspectives.

Purpose and Overview

The purpose of this book is to describe, assess, and critique biomedicalization and its influence as a larger social trend on the health field and specifically in the area of alcohol research, policy, and programs. Chapter 2 will provide the larger context of this pervasive phenomenon termed biomedicalization and will describe how this process developed and has been experienced and expressed in the diverse fields of aging, psychiatry/mental health, and women's health. In each field, the focus has been on pathologizing deviant and sometimes normal states, introducing technology primarily in the area of pharmacology as a means to treat this condition or disease, and a recurring emphasis on the "problem" being individualized and hence resulting in a limited purview which effectively eliminates from consideration social or environmental influences.

Chapter 3 will introduce the area of alcohol studies as a specific, in-depth case study. In order to understand the context and consequences of biomedicalization in this field, it is important to understand the historical conditions under which biomedicalization has come to dominate and flourish. This chapter initially describes the historical process of the medicalization of the alcohol field. First, the political and economic conditions in the late nineteenth and early twentieth centuries that led to an "addiction" ideology and to the subsequent medicalization of social problems are traced. Included in this discussion is the rise of Alcoholics Anonymous (AA) and other self-help groups that, along with professional groups such as the American Medical Association, worked hard and successfully to define alcoholism as a disease. Next, I examine how the assumptions behind the disease concept of alcoholism continue to dominate current work in the alcohol field. The last part of the chapter focuses on a number of historical "outgrowths" of the AA movement including the twelve-step philosophy that has rapidly proliferated into other areas of life affecting a much larger range of people, for example, Al-Anon, Adult Children of Alcoholics (ACAs) which, in turn, has led to the emergence of the diffuse and broad co-dependency movement.

In the fourth chapter, the process by which NIAAA, along with the National Institute on Drug Abuse (NIDA) and the National Institute of Mental Health (NIMH) moved their research mandates from the Alcohol Drug Abuse and Mental Health Administration (ADAMHA) to the National Institutes of Health (NIH) in 1992 is delineated. By examining Senate and House hearings during that time, the rationale for this move from ADAMHA to NIH is explored as well as some of the concerns voiced by various legislators, and by other individuals, about the possible problems posed by such an organizational and ideological shift. It is argued that the move to NIH codified, legitimated, and intensified NIAAA's primary focus on the "disease" model of alcoholism. This move to NIH enhanced NIAAA's framing of alcohol problems as a disease and thus moved it further away from a broader range of important issues in the alcohol field. It also further enhanced the biomedicalization of the alcohol field.

Chapter 5 provides multiple examples of how NIAAA's movement to NIH crystallized NIAAA's organizational and political identity as a biomedical agency. This process is shown through an examination of documents including NIAAA's Strategic Plan; by an assessment of grant funding by branch or division from 1990-2003; and through a content analysis of the 10 NIAAA *Reports to Congress* (1971-2000). I also discuss some recent changes at NIAAA including its reorganization and its defunding of the ETOH database which directly relate to an increased biomedical focus. Each of these areas of inquiry demonstrates how NIAAA's priorities have increasingly moved towards biomedicalizing the alcohol field. As a consequence of this shift, NIAAA's attention to social and environmental concerns, while not absent, has been steadily reduced and minimized.

In chapter 6, I provide an international example of the direction of alcohol research by assessing how Sweden has handled alcohol problems and currently funds alcohol research. This work was done while I was on sabbatical in Stockholm and had the privilege of being a Fulbright Scholar in the spring of 2004. Sweden, the welfare state, has a strong history of social medicine in which environmental and social processes are considered in the context of medicine. Moreover, Sweden also historically has addressed social problems from a community perspective. Thus, a comparison between Sweden and the U.S. of how alcohol issues are handled and funded is extremely important in beginning to address whether biomedicalization in the alcohol field is a global phenomenon.

The concluding chapter outlines the rationale for an expanded discourse between social scientists and biomedical researchers working on social problems, particularly alcohol issues. Unfortunately, social scientists have not been very successful in demonstrating their important contributions to many fields, including the alcohol field. First, four areas in which social science has clearly been important in conceptualizing and defining the issues, conducting the research, and developing policies are presented. These areas include: (1) alcohol epidemiology and specifically the focus on drinking patterns; (2) alcohol's contribution to the burden of disease; (3) assessment of the impact of alcohol control policies including taxation, alcohol outlet density, and college drinking policies; and (4) the benefits of alcohol screening instruments and brief interventions for the identification of at-risk drinkers as well as individuals with alcohol dependence. Finally, the ways in which biomedical and social science research in the alcohol field and other fields can work together to go beyond a too common mutually dismissive co-existence are discussed in the hope that by working in an interdisciplinary fashion, we can continue to define important areas of research, practice, and treatment agendas that blend individual and environmental factors.

My hope is that this book will stimulate more discussion of the processes by which social problems, and specifically alcohol issues, are framed, managed, and studied. As someone who has been working in the alcohol field for more than thirty years, originally as a clinician and then as a researcher and teacher, I am writing this book from the perspective of someone trained in public health who looks at the alcohol field as a social scientist. Most of the research I have conducted throughout my career has been epidemiological and methodological. Over the years, I have questioned why some types of research are encouraged and funded, and why other forms of inquiry are very much discouraged. The ongoing multiple efforts to biomedicalize alcohol research is not just a matter of funding biomedical research, it is also a movement to reframe fundamentally social science research so that it resembles or becomes "harder science." The goal of this book is not to argue the importance of one model or another. Rather, this book seeks to begin an ongoing and open dialogue concerning why and how social scientists and biomedical researchers can, and I will argue, must work together to unravel the complexities of alcohol related problems.

References

APA. (1994). *Diagnostic and Statistical Manual of Mental Disorders (DSM-IV)*. Washington, D.C.

Babor, T. F. (2000). Talk is cheap: Measuring drinking outcomes in clinical trials. *Journal of Studies on Alcohol, 61*, 55-63.

Clarke, A. E., Shim, J. K., Mamo, L., Fosket, J. R. & Fishman, J. R. (2003). Biomedicalization: Technoscientific transformations of health, illness, and U.S. biomedicine. *American Sociological Review, 68*, 161-194.

Conrad, P. (1999). Introduction: Sociological perspectives on the New Genetics: An overview. In *Sociological Perspectives on the New Genetics* (pp. 1-12). Malden, MA: Blackwell Publishers.

Conrad, P., & Schneider, J. W. (1980). *Deviance and medicalization: From badness to sickness*. St. Louis, MO.: Moseby Press.

Midanik, L. T. (2004). Biomedicalization and alcohol studies: Implications for policy. *Journal of Public Health Policy, 25*, 211-228.

Morgan, P. A. (1988). Power, politics and public health: The political power of the alcohol beverage industry. *Journal of Public Health Policy, 9*, 177-197.

Moynihan, R., Heath, I. & Henry, D. (2002). Selling sickness: The pharmaceutical industry and disease mongering. *British Medical Journal, 324*(7342), 886-891.

Siegler, M., Osmond, H., & Newell, S. (1968). Models of Alcoholism. *Journal of Studies on Alcohol, 29*(3), 571-591.

2

Biomedicalization: At-Risk as a Steady State

> *"... health itself and the proper management of chronic illnesses are becoming individual moral responsibilities to be fulfilled through improved access to knowledge, self-surveillance, prevention, risk assessment, and the treatment of risk, and the consumption of appropriate self-help/biomedical goods and services."*
> *(Clarke, Shim, Mamo, Fosket & Fishman, 2003, p. 162)*

Introduction

Biomedicalization is seen by many as the natural outgrowth of continued scientific progress—a movement towards improving the quality and quantity of life through scientific inquiries using biomedical perspectives and methods. This approach carries with it the assumption that with "proper" risk assessment, detection, and treatment, our lives can be lengthened, improved, and indeed, more fulfilling. Yet, critics of this movement with its emphasis on biotechnology and genetic research question its ability to deliver on these promises, its dominance of all other perspectives, and its inherent assumption that this approach is both desirable and needed without considering the potential consequences, ethical and otherwise (Duster, 2003; Nelkin & Tancredi, 1994). There is also concern about how biomedicalization can change our traditional concepts of health as we discover more conditions for which we are at-risk and, as the above quote suggests, health maintenance is seen as the responsibility of the individual. Despite these concerns, there is too little debate concerning the directions for future inquiry and policy.

The purpose of this chapter is to lay the groundwork for an historical understanding of how medicalization and biomedicalization have developed. First, a brief history of the profession of medicine in the U.S. will be discussed in order to understand the growth of medicine and its "professional dominance" (Friedson, 1970). A discussion of the consequences of the progressive medicalization of deviance will follow. Third, the process of biomedicalization will be assessed from both a conceptual and definitional viewpoint. Its position either as an extension of medicalization or as something distinct from medicalization will be discussed. Within the discussion of biomedicalization, the impact of the New Genetics and its influence on the biomedicalization process will also be explored. Finally, specific case examples of biomedicalization will be presented that include both deviant and stigmatized behaviors, e.g., mental illness, drug addiction, as well as natural life course events such as menopause, reproductive health, and aging. These case examples will emphasize how biomedicalization has differentially occurred across various disciplines.

The Medicalization of Social Problems

Medicine as the Dominant Profession

In order to appreciate fully the historical shift from viewing deviant behavior as bad behavior to sick or diseased behavior, it is first necessary to understand the development of medicine in the U.S. Several excellent books have been written on the history of medicine and its eventual dominance as a profession (Brown, 1979; Friedson, 1970; Starr, 1982; Stevens, 1971). Various frameworks have been developed to understand the progression of medicine in the U.S. and the political and economic forces that shaped the way medicine is practiced today. This section will briefly discuss the development of medicine, and hence the biomedical model, as a dominant force in the U.S.

Prior to 1850, during a period Starr (1982) called "Medicine in the Democratic Culture," medicine was not formally organized. One explanation for this is the on-going tension that existed in Colonial America between the concept of a democratic culture where there was a belief that anyone could do anything they wanted and a view that oversight needed to be imposed by some groups. Early attempts of physicians to license themselves were generally not successful at least partly due to the general feeling that governance over a profession belongs to the population at-large. This is well expressed in 1844 by Oliver Wendell Holmes in his speech to the graduating class at Harvard University Medical School: "...no profession was allowed to be the best judge of its own men and doctrines" (cited in Starr, 1982, p. 30)

From this early American democratic culture, three strands of health care emerged:

1. *Domestic Medicine* as practiced in the home by predominantly women caregivers;
2. *Lay Healers* consisting of midwives, botanic practitioners, and natural bonesetters; and
3. *Professional Medicine* defined as trained physicians.

While all three spheres co-existed in colonial America, professional medicine was, in many ways, the most threatening to the democratic society given its early dedication to professionalism that was seen as the antithesis of an open society. As a result, physicians could not obtain the power needed to dictate its authoritative views in society. Further, unlike in England where physicians were part of the privileged class, in colonial America the boundary between profession and trade was unclear. All types of people became doctors; there were no consistent standards. Physicians often had to do other work to supplement their income; moreover, training varied widely. Most physicians were apprenticed; a few were trained in European universities. When medical societies did form, they could not regulate or license physicians. If they imposed stringent criteria, they risked losing their membership (Starr, 1982).

However, the role of physicians did change after 1850 during which time medicine began to organize seriously and eventually develop its authority over itself and other health-related professions. Several forces contributed to medicine's professionalization. First, indirect costs for medical care, e.g., transportation and supplies, which had been quite high relative to wages, dropped considerably due to the invention and widespread use of the automobile and telephone. Telephones first became available in the late 1870s and initially connected physicians with drugstores that served as message centers. Automobiles were invented in the1890s but became more reliable between 1906-1912, allowing physicians to see more patients thereby increasing their income. Thus, the profession itself became much more attractive to prospective recruits.Second, the U.S. was experiencing a growing urbanization with greater separation between work and residence. There were fewer people at home to care for sick family members, and thus, hospitals run and staffed by physicians developed as centralized places of care. Home care was replaced by the hospital (outpatient and inpatient care) as the locus for treatment. Third, in 1847, the American Medical Association (AMA) was founded by a group of younger, more discontented physicians who were basically reacting to the continual failure of the states to regulate and license medical practice. The AMA had a clear agenda that included raising and standardizing the requirements for medical degrees in order to keep out irregular practitioners (anyone other than physicians). It also developed a Code of Ethics to foster some level of professional accountability and solidarity (AMA, 1847).

A fourth and very powerful force that medicine used to consolidate its authority was the control over medical education (licensure, regulation, admissions to medical schools, etc.). Early on in U.S. medical schools, faculty collected fees directly from students if they passed exams. Therefore, controlling the quality of student that remained in the school was quite difficult because of this financial arrangement. This lack of standardization led the AMA to commission the Carnegie Foundation in the early 1900s to conduct an evaluation of medical schools in the U.S. and Canada. The result was a report written by Abraham Flexner in 1910 entitled *Medical Education in the United States and Canada. A Report to the Carnegie Foundation for the Advancement of Teaching*. This evaluation later became known as the Flexner Report (Flexner, 1910). In the comprehensive report of the 150 medical schools in the U.S. and Canada, Flexner summarized the standards or lack thereof in these medical schools. The Flexner Report uncovered fraudulent claims of some medical schools and argued that only five schools of medicine met any reasonable standards: Harvard, Johns Hopkins, Case Western Reserve, University of Toronto, and McGill. As a result of the Flexner Report, most medical schools were unable to meet the imposed standards that included a strong curriculum in both the classroom and in the laboratory.

A fifth factor that led to the professional dominance of medicine was its eventual control of the pharmaceutical industry. Prior to 1914, there was no official federal drug policy. Among the drugs that were available and accessible were patent medicines, composed essentially of opiates. A major impetus for this reform was the passage of the federal Food and Drug Act in 1906 following the publication of *The Jungle* (Sinclair, 1906) earlier in the year that exposed the unhealthy practices of the meat packing industry in Chicago. The AMA, as an organization, supported muckraking journalists who exposed fraudulent claims particularly of patent medicines. They also acquired resources to create their own regulatory apparatus. Drug makers were eventually forced to recognize that they would need doctors to "market" their drugs; hence, with the passage of the Harrison Act of 1914, that limited the distribution of opiates to physicians and later was interpreted by a series of Supreme Court decisions as a prohibition of opiate prescriptions, the medicalization of the pharmaceutical industry was accomplished (Starr, 1982).

The sixth force that served to consolidate the professional authority of medicine was its successful moral crusading in areas that have been termed social reform, for example, the outlawing of abortion. Abortion was a fairly common medical procedure prior to the 1860s, usually done by physicians and midwives. A pregnancy was not considered viable until "quickening" occurred (first perception of fetal movement—usually in the second trimester). Thus, it was not defined as a social, moral, or medical problem at that time. Conrad and Schneider (2001) make a strong argument that a series of circumstances beginning in 1840 contributed to the shift from abortion being seen as a routine medical procedure to a moral crusade in which abortion was seen as an attack on the lives of children. First, after 1840, abortion became more public which increased advertising of abortion clinics in the media. This suggests that an increasing number of women were obtaining abortions at that time. Second, much of the increase in use of abortion services was attributed to white, Protestant, native-born, middle and upper-class women, whereas prior to this time, abortion services were primarily used by "lower-class" women in "unfortunate circumstances." This shift in clientele raised greater fears among middle and upper-class men (including physicians and legislators) who were concerned about the possibility of disproportionately higher birthrates among immigrants to the U.S. Control of abortion had the potential of increasing the birthrate of native American children, thereby ensuring that the proportion of native births to immigrant births was "right" for America (Conrad & Schneider, 2001).

In terms of medical dominance, Conrad and Schneider (2001) contend that the abortion crusade by physicians was yet another strategy for ensuring that non-physician practitioners who commonly performed abortions, and in many cases had lucrative practices, would no longer be able to do so. Outlawing abortion was another mechanism for asserting medical dominance.

In addition to the forces just discussed to consolidate professional dominance for the medical profession, scientific events and discoveries also occurred that served to reinforce medicine's importance in American culture. For example, anesthesia and antisepsis were developed which greatly aided surgeons and generally improved hospital care. In addition, the "germ theory of disease" gained prominence led to the "doctrine of specific ideology," in which the assumption was that for every disease there was a specific cause (germ or agent). The focus of disease was the body with much less emphasis on environmental factors (Tesh, 1988).

That medicine became the dominant profession that established a monopoly over medical practice is not disputed. It was no doubt unimaginable for Friedson writing in 1970 to consider a reversal of medicine's prestige. However, McKinlay and Stoeckle (1997) provide a very strong argument that this process may in fact be reversing to some extent through a process of "proletarianization" that occurs when "...an occupational category is divested of control over certain prerogatives relating to the location, content, and essentiality of its task activities, thereby subordinating it to the broader requirements of production under advanced capitalism" (McKinlay & Stoeckle, 2001, p. 189). They argue that there have been substantial legal, economic, and social shifts in medicine that have decreased its professional dominance. These shifts include a change in criteria for entrance into medical schools with more pressure (fiscal and legal) to recruit and admit women and minorities; more authority by the federal government to dictate curricula in medical schools due to mandates contained in federal training programs, etc.; less autonomy over the terms and content of physicians' work given newer managed care arrangements; less "ownership" of patients given that patients are often considered clients of an organization as opposed to being patients of an individual physician; less or no ownership of the "physical plant" and the technology which is mainly owned by the organization; and fewer fee-for-service arrangements and increasingly more salaried physicians. With the advent and growth of managed care nationwide, one can only assume that this pattern will continue. Yet, even as the medical profession itself is undergoing these changes, the medical model has continued to have a strong influence in American medicine.

Medicalization of Deviance

These events each in their own way contributed to medicine's dominance of other professions as well as to the strong belief that medicine had the power, or at least the potential, to cure all ills – including what were once considered to be social ills. As medicine accumulated "wins" over infectious diseases, the idea that disease was an individualized problem contained within a human body gained widespread acceptance. Larger frameworks such as environmental, cultural, social, or economic factors were generally not included in this medical framework. Thus, as treatment rather than punishment emerged as the preferred strategy for certain types of deviance, the natural next step was to place this deviance under the umbrella of the medical model, thus shifting the definition of a deviant person from someone who engaged in "bad" behavior to someone who was "sick" and in need of medical treatment.

The literature on the history of medicalization recognizes the work of Talcott Parsons (1975), Irving Zola (1972; 1975), and Elliot Friedson (1970) among the key authors who critically examined medicine as an agent of social control. One critic of the expanded role of medicine even went so far as to argue that the promise of medical advances rendered any form of "non-perfect" health as abnormal or deviant and, in fact, the treatment for ill health had itself become iatrogenic (Illich, 1975). Each of these authors questioned either explicitly or implicitly the expansion of the medical model to a wider range of behaviors and conditions previously considered social, moral, and certainly not medical. They expressed concern about the trajectory of increased medicalization and its consequences as more behaviors become labelled as deviant or pathological and thus become medical problems. Along with these concerns is the fear that as the medical profession becomes more expansive, it will assume an even more powerful role over our lives. While much has been written about medicalization since these early studies, the seminal work that focuses on the medicalization of deviance is Peter Conrad and Joseph Schneider's book *Deviance and Medicalization. From Badness to Sickness* written in 1980 and updated in 1992. Much of the following discussion on the definition and conceptualization of medicalization stems from this book as well as from more recent work by Peter Conrad.

Medicalization has traditionally been defined as the process whereby there is an expansion of the medical model to encompass behaviors previously defined as non-medical and to claim them as "medical problems" rendering the "problem" in need of "medical treatment." This expansion of medical control has been observed in such diverse areas as hyperactive children, suicide, childbirth, aging, crime, mental retardation, mental illness, alcoholism, drug addiction, child abuse, learning problems, impotence, and menopause. That these areas have been designated as medical problems is not necessarily negative; yet, to the extent that these areas have become medicalized to the exclusion of other ways of defining and addressing these issues is potentially a major problem.

The ways in which behaviors become medicalized is extremely important to consider in that it provides a framework to assess how the process works and with which behaviors. Conrad and Schneider (1980) delineated a five-stage sequential model that illustrates how deviant behaviors become medicalized. These five stages include:

1.   definition of behavior as deviant;
2.   prospecting: medical discovery;
3.   claims-making: medical and nonmedical interests;
4.   legitimacy: securing medical turf; and
5.   institutionalization of a medical deviance designation. It is notable that few models exist that specifically address this process, and as of 1992, there has been little evaluation to test this specific model (Conrad, 1992).

Conrad and Schneider's model is innovative and helpful for illustrating the theoretical processes by which definitional shifts occur as bad behaviors become defined as illnesses or sickness. The sequential stages of the model imply, by design, that one stage follows the next in a specific order. This may or may not be the case for specific areas of deviancy particularly as new pharmaceutical products are discovered, more off label uses are promoted, and the pharmaceutical industry makes strong efforts to expand its market by medicalizing behaviors. For example, Moynihan (2003) chronicles the pharmaceutical's active interest, sponsorship, and promotion of medicalizing female sexual dysfunction on par with erectile dysfunction for which drugs such as sildenafil (Viagara) yielded Pfizer 1.5 billion in 2001. Through a series of pharmaceutical sponsored meetings occurring beginning in 1997 that included industry representatives as well as scholars from schools of medicine, movement was made towards the recognition and adoption of a common ways to assess and diagnose this disorder in both males and females. Moynihan (2003) also points out that two of the three authors of the study that found that 43 percent of women over the age of eighteen experience sexual dysfunction "...disclosed close links to Pfizer" (Moynihan, 2003, p. 47). From this example, one could argue that the discovery of this female sexual dysfunction was not necessarily "medical" per se but rather the result of a confluence of forces, primarily biotechnical, which were embraced by the medical profession and claimed as their own. Thus, without the development of newer and better pharmaceuticals treatments, specific behaviors may not have been considered abnormal or deviant but became that way to drive market forces.

In fact, the role of the pharmaceutical industry in the medicalization of specific behaviors or conditions is powerful in shaping how deviance is defined. Waddington (2000) provides a compelling example in the area of sports with the advent of sports medicine. He defines medicalization as "...the development of a ideology justifying increasing medical intervention" (p. 121). While disagreeing with the reasoning that widespread drug use in sports is solely due to increased availability of drugs such as steroids and amphetamines after World War II (Donahoe & Johnson, 1986; Verroken, 1996), he points to the combination of the access of drugs to athletes by the pharmaceutical industry as well as the social context of sports such as its growing competitiveness and its political ramifications (Waddington, 2000). Thus, this sequential model which presupposes that a behavior is labelled as deviant before a medical or pharmaceutical discovery is made may be more appropriate for some forms of medicalization of deviance in which behaviors were already viewed as "bad," for example, mental illness, and less appropriate for contemporary "life-stage" deviance such as aging or menopause.

While Conrad and Schneider's model is extremely important to begin the process of explaining how medicalization occurs, we are left to ponder if there are examples of deviant behavior that have not been medicalized and, if so, what does this mean for the individual who exhibits this behavior, medical treatment personnel, institutions who are responsible for individuals who exhibit this deviant behavior, and society? What can we learn from those behaviors that have not been labelled as deviant by some institutional force?

Alternatively, should we be demedicalizing, or partially demedicalizing, some behaviors and if so, which ones and how would this be done? What would replace the medical profession as the dominant institutional force? While it is unfair to expect that Conrad and Schneider's model can address all of these questions, they are important to consider as we move the discussion to the advantages and disadvantages of the medicalization of deviance.

While medical sociologists have been critical of the medicalization process for the ways in which it has impacted those labelled as deviant and the ways it has contributed to enlarging the sphere of medicine, one cannot conclude that medicalization itself is necessarily bad or undesirable. While it might be easy to focus solely on the negative aspects of medicalizing deviant behavior, it is important to consider that this process occurs in a larger context within multiple systems, most notably the criminal justice and the health care systems. As such, it is important to consider both the benefits and problems associated with this process.

In the 1992 edition of their work, Conrad and Schneider (1992) list six areas that can be considered the "brighter side" of the social consequences of medicalization. Specifically, these advantages include: "…the creation of humanitarian and non-punitive sanctions; the extension of the sick role to some deviants; a reduction of individual responsibility, blame, and possibly stigma for deviance; an optimistic therapeutic ideology; care and treatment rendered by a prestigious medical profession; and, the availability of a more flexible and often more efficient means of social control" (pp. 246-248).

These more positive aspects of medicalization are important for considering why such a process occurs. Humanitarian efforts have often been used to explain why shifts have occurred between classification of sinful behaviors and "diseased" behaviors. There is a long history of this movement for "the good of the patient" particularly in the area of mental illness. Medical diagnoses appear to be welcomed by consumer groups such as the National Alliance for the Mentally Ill (NAMI) as a means of decreasing shame and guilt and reducing social stigma when presenting the condition to the public at-large (Conrad, 2004). As discussed extensively by Conrad and Schneider (1992) the role of individual responsibility in "sick role" behavior weighs heavily on the shift from seeing a specific behavior as illegal or immoral to one that is a symptom of a much larger illness. In the alcohol field, the extent to which excessive drinking is attributed to individual responsibility is passionately debated in the literature with the disease model on one extreme making the case that alcoholism is not the "fault" of the individual, and the other side arguing that "alcoholism as a disease" is a myth and that the main focus in this area should be on individual responsibility (Fingarette, 1988).

Reliance on "individual responsibility" has also been used to provide or deny benefits to those diagnosed as alcoholic. For example, in 1988, the Supreme Court decided for the Veterans Administration who refused to grant two veterans, who were recovering alcoholics, extensions of time to use their veterans' educational benefits (*Eugene Traynor, Petitioner v. Thomas K. Turnage, Administrator, Veterans'Administation and The Veterans' Administration*, 1988; *James P. McKelvey, Petitioner v. Thomas K. Turnage, Administrator of Veterans' Affairs, et al.*, 1988) Citing that federal programs cannot discriminate against individuals who are handicapped, the plaintiffs argued that alcoholism is a disease, that they are disabled by virtue of their alcoholism, and therefore, they should be exempt from the designated time limits and allowed to utilize their educational benefits. The Supreme Court, however, in a 4 to 3 vote argued that alcoholism included a component of "willful misconduct" which therefore placed the responsibility on the plaintiffs for not utilizing their educational benefits during the time limit of ten years after they left the service. This was a significant decision even though the Supreme Court refused to formally define alcoholism per se. The decision, did however, differentiate between disabilities in which willful misconduct is not involved and behaviors which include the "willful" involvement of the individual (Taylor, 1988).

All of these possible benefits of medicalization discussed by Conrad and Schneider (1992) provide useful explanations for why this process appears to be accelerating over time. Yet, it is important to consider if some advantages of medicalization are more important than others at different points in times. In other words, do humanitarian reasons represent a stronger driving force for medicalizing deviancy than, for example, the flexibility of the medical model and does the relative importance of each of these factors differ by which behavior is medicalized? Because medicalization, as defined, heavily focuses on the "medical" profession *per se*, other allied groups that can better be described as biomedical or biotechnical, are often placed under the rubric of "medical" and thus their independent contributions to the expansion of the medical model or other models is to some extent masked. For this reason, the privatization of biomedical "discoveries" and therefore the financial profitability of medicalization is often not included in the analysis, and yet it can be a very important consequence in the less regulated medical and health market as is the case in the U.S. (Gambrill, in press). Privitization and the profit motive describes the nature of the health care system as well as the direction of institutions within the system and will be explored more fully under biomedicalization. The pharmaceutical industry has been called "disease mongers" in their medicalization of every day life (Moynihan, Heath & Henry, 2002). Moynihan et al. (2002) point to the excessive profits made by the pharmaceutical industry in their expansion of markets to healthy people by making them feel they are ill. The promotion of baldness and bone loss due to aging as "serious diseases" leaves consumers feeling that these conditions are "...widespread, serious, and treatable" (p. 886). Other ways of handling these issues are largely ignored thus moving the public towards taking medications—a profitable direction for the pharmaceutical industry.

The "darker" sides, or disadvantages, of medicalization as explicated by Conrad and Schneider (1992) appear to be quite extensive. These include:

1. dislocation of responsibility;
2. assumption of the moral neutrality of medicine;
3. domination of expert control;
4. medical social control;
5. individualization of social problems;
6. depoliticization of deviant behavior; and
7. exclusion of evil (Conrad & Schneider, 1992, pp. 248-252).

As discussed above, just how much responsibility is attributed to the individual for his or her condition is extremely problematic. Even with illnesses that are less controversial and therefore not considered deviant *per se* such as diabetes or hypertension, individual-level behavior does contribute to how well the disease is controlled and that, in turn, is influenced by social forces.

Like the issue of individual responsibility that can act as both an advantage and disadvantage to medicalization depending upon the circumstances, such is the case for the corporate profit motive associated with medicalization. From the vantage point of the medical profession and the institutions aligned with it, the increase in patients to treat (because of medicalization) may or may not be profit-able. In traditional fee-for-service medical practices, adding more insured patients to ones caseload or adding patients who could pay out of pocket for health care would directly increase income to individual physicians with additional incentives to see them more often and include, in addition, expensive medical tests and procedures. In managed care arrangements in which the institutional goals are to lower costs by decreasing medical visits and unnecessary medical tests and procedures, this would result in only marginal increases in income from new patients or tests. Yet, health services research in the area of alcohol and drug abuse, for example, indicates that providing treatment for these "behaviors" or conditions does lower over-all inappropriate usage of health services in managed care organizations (Armstrong, Midanik, Klatsky & Lazere, 2001). This provides not only support for humanitarian reasons for providing care, but also for financial incentives to further medicalize behaviors as they affect overall use of health services.

While the model of medicalization and its advantages and disadvantages as outlined in Conrad and Schneider's book are extremely useful in assessing and understanding the process, the current health climate in which there has been substantial increases in privatization, biotechnology, and financial incentives for medical discoveries have all necessitated moving beyond the landscape of medicalization. Biomedicine and biotechnology fuels a newer group of vested interests that have moved medicalization beyond its original definition.

By framing deviance as a medical issue, Conrad and Schneider (1992) list several advantages. First, the problem is framed in a humanitarian fashion as opposed to one of sin or evil. Thus it allows for the deviant to be seen as one who is treatable and thus more worthy. Second, medicalizing social problems allows the deviant to assume the sick role and derive its benefits. The deviant is not seen as the cause of his or her condition but rather an unfortunate, yet legitimately sick individual who happens to have an illness but is not responsible for it. Third, because treatment is available for this condition, there is an optimism concerning the outcome of this illness. By framing deviancy as an illness, it becomes aligned with other illnesses that have been successfully treated, for example, polio, through scientific research and medical interventions. The very act of treating an illness brings with it an element of hope for cure. Fourth, the involvement of the medical profession in the definition, diagnosis and treatment of deviancy brings with it a prestige that could not be achieved otherwise. The dominance of medicine in the hierarchy of professions provides higher status to whatever disease it attempts to conquer. Finally, the medicalization of deviance provides a more flexible means to control it. Unlike legal or judicial procedures, medical treatments can be given on a more informal basis and can include a wider array of choices to meet individual needs.

There are, however, several disadvantages to controlling deviance by the use of a medical framework. First, by eliminating individual responsibility from the development of an illness, a "two-class citizenship" (Conrad and Schneider, 1992, p. 249) is created: those who are responsible for their actions and those that are not with the latter having a lower status. Second, the medical model is often defined as morally neutral, devoid of moral judgment. This neutrality includes an emphasis on objectivity, rationality, and science. Yet Conrad and Schneider (1992) clearly point out, "...medical designations *are* social judgments, and the adoption of a medical model of behavior, a political decision" (p. 35). By assuming a value-free stance, important political arenas that directly impact how deviance is defined are ignored. For example, homosexuality was considered originally as a condition under the category of Sociopathic Personality Disturbance in the first edition of the DSM in1952. By 1968 when the second edition of DMS was published, homosexuality was listed under Sexual Deviation, a category under Personality Disorders and Certain Other Non-Psychotic Mental Disorders. After much political pressure by gay activists, a revised diagnosis of "Ego-Dystonic Homosexuality" replaced the earlier one but it was not until 1987, with the publication of the DSM-III-R that homosexuality as a diagnosis was completely eliminated (Kirk & Kutchins, 1992). Third, the authority of the medical profession allows it to reign as experts in the areas they pursue – including deviance. While other professionals may have more expertise, when a problem is medicalized, the medical profession becomes the predominant voice (Kirk & Kutchins, 1992). Fourth, while a broader array of treatments may be available under the purview of medicine, they can be effectively used as objects of social control and can go beyond what could be done in other areas. For example, severe treatments, such as psychosurgery and powerful drugs, can be given to patients defined as deviant as a way to align their behaviors with the status quo.

Fifth, when deviance is medicalized it assumes an individualistic perspective. Thus, alcoholism, juvenile delinquency, and other negative behaviors are defined on the individual level only; other levels of analysis are less often considered. By diminishing the contributions of other perspectives such as family, community, and society, treatment is focused on changing the individual. In circumstances where the individual is either unwilling or unable to change, there may be a tendency to "blame the victim" as opposed to exploring other alternative ways to handle the problem. By medicalizing alcoholism, the effect of larger, broader policies such as marketing efforts by the alcohol beverage industry aimed towards youth and people of color, alcohol beverage taxation, and packaging of malt liquor in large, 40-ounce containers are less likely to be considered serious or important strategies. Finally, when deviance is medicalized, there is little if any emphasis on the politics of such a process. By declaring an individual or a group of individuals deviant by giving their behavior a diagnosis and offering a pharmaceutical solution, the social and political context of this deviance is lost and thus depoliticized . Dissenters can then be viewed as deviants and treated accordingly without regard for the motives behind their behaviors. Thus, virtually any threat to the status quo can be handled through a "legitimate" medical process that, in turn, devalues the political and emphasizes the individual.

Over time, medical treatments for deviant behavior have been considered progress for modern society (Conrad and Schneider, 1992). This progress has taken many forms including different types of therapy (as diverse as psychotherapy and shock therapy), surgery, and psychotrophic medications. Regardless of what type of treatment is used, the key factor in the medicalization of deviancy is that it becomes socially defined as a medical problem, requiring a medical diagnosis and treatment model. Even when genetics links are not determined and biomedical solutions are not forthcoming, the use of medical and scientific language in connection with the behavior or condition assures the public that progress is being made (Boyle, 2002).

This practice of medicalizing a wide range of behaviors and declaring them diseases led Dr. Richard Smith, editor of the *British Medical Journal,* to publish a humorous yet poignant clinical review entitled "In Search of 'Non-Disease'"—a report on a vote conducted on the internet to determine the top non-diseases as a method of illustrating how difficult it is to define disease (Smith, 2002). Based on this non-random survey, the top twenty non-diseases (in descending order) were: "ageing, work, boredom, bags under eyes, ignorance, baldness, freckles, big ears, grey or white hair, ugliness, childbirth, allergy to the 21st century, jet lag, unhappiness, cellulite, hangover, anxiety about penis size/envy, pregnancy, road rage and loneliness" (p. 4). Smith's (2002) results indicate more seriously how far the medical and biomedical models have come. While his survey shows aging as a non-disease, Estes and Binney (1989) argue quite differently (more fully discussed below) and others have argued that pregnancy and childbirth have indeed become diseases as well. Could ugliness or boredom be lagging far behind? We await the next version of the DSM to find out.

Biomedicalization

Biomedicalization has been conceptualized and defined in several ways depending on one's level of analysis, one's position or point of entry into the discourse, and the discipline or field from which one comes. For some, the focus is in one area and the writing reflects the authors' basic disagreement and frustration with how an entire field is being viewed within the research community as well as the general public, for example, biomedicalization in the aging field (Estes & Binney, 1989; Lyman, 1989). For others, the focus is more general and attention is placed on the larger social trends affecting health care and health care service delivery in the U.S. (Clarke et al., 2003). Clarke et al.'s (2003) definition of biomedicalization is very broad in scope and describes this process from the standpoint of the shifts in American medicine that have emerged during the last two decades. Central to their concept of biomedicalization is essential importance of technoscience that includes "…innovations as molecular biology, biotechnologies, genomization, transplant medicine, and new medical technologies" (p. 162). They argue that these new areas are not only shaping the current medical arena, they are also extending the boundaries of medicalization and as such are changing the ways in which the health system operates. As a result, medicalization has moved beyond defining increasingly more areas of deviancy or natural life stages as medical problems, but has, in fact, become a new process, biomedicalization, which has redefined the way health itself is conceptualized. As the quote from Clarke et al. (2003) suggests at the beginning of this chapter, this shift from medicalization to biomedicalization represents how health itself is defined. They also suggest that this process began around 1985 when the shift moved from an attempt to control the external environment to one that focused primarily on controlling biological or internal processes through scientific discovery and other biotechnical applications.

Clarke et al. (2003) argue that there are five main processes of biomedicalization that have shaped and framed biomedicalization. These include major political and economic shifts that have allowed technological advancement to be "corporatized and privatized," thus the research that leads to these products as well as the marketing of these products does not rely on public funding and is thus far less accountable to the public. A second process is the focus on health, risk and surveillance that has led to the wide-spread belief that the potential for illness is within everyone and that the role of medicine is to discover increasingly more risk factors to prevent possible future disease. The third process, termed "technoscientization of biomedicine" is defined as the ways in which technology has changed how biomedical information is organized, developed and distributed. The fourth process focuses on dissemination of information on health through channels such as the Internet, traditional and non-traditional forms of advertising, and newspaper and other media outlets. The final process is a philosophical shift from medicine's focus on "control" of bodies to one of transforming them so that they can be fundamentally changed in a presumably positive direction. This last process represents perhaps the most seductive feature of biomedicalization, because it includes the hope, maybe the promise, and the possible opportunity for change as well as for new and healthier identities.

Contemporary discussion of this new biomedical emphasis has been emerging primarily but not exclusively in medicine. For example, Hewa and Hetherington (1995) have explored why this mechanistic, biomedical model became dominant in the medical field and continues to reign despite its obvious limitations. Using Max Weber's theory of rationalization, they trace the roots of this model to the writings of René Descartes and William Harvey in the late sixteenth century following the Reformation (Hewa & Hetherington, 1995). In these works, the human body, as separate from the mind, worked like a machine. Illness or disease was then described in terms of "mechanical breakdown"—as an inability for parts of the body to function. This focus on the internal workings of the human body to define disease and to determine interventions supported the importance of calculability and predictability the idea of as crucial for medical practice. Moreover, because the biomedical model does not take into account larger psychosocial issues, it reduces the definition of health to the absence of disease (Engel, 1977; 1980). Moreover, to some extent, one is never fully health nor diseased but instead is somewhat inbetween: needing on-going medical intervention to monitor varying levels of risk.

While Engel's (1977; 1980) seminal works on the importance of using a biopsychosocial approach in medicine can be seen as a call for a much larger definition of health and disease, it seems to have had little impact on how medicine is practiced today. Hewa and Hetherington (1995) propose that it is the "predictability, calculability, and control" so valued in our society and so closely associated with the biomedical model that continues to ensure its domination over other models and particularly over the broader biopsychosocial approach. Hawkins (1994) takes this argument even further to observe that "...it is both sad and ironic that the biopsychosocial paradigm that Engel a quarter of a century ago saw as remedying the deficiencies of the biomedical model should now appear—in one of those Kuhnian "gestalt switches"—as firmly allied with alternative medicine" (Hawkins, 1994, p. 56).

In a very real sense, Clarke et al.'s (2003) perspective represents a top-down approach in which the level of analysis is systems and institutions. However, other definitions of biomedicalization have emerged that are more particularistic, emanating from observations within specific disciplines or areas of research. Estes and Binney (1989), perhaps the first ones to refer to biomedicalization in the late 1980s, but they were less concerned about the technological aspects of biomedicalization, although they acknowledge its influence, and more concerned about the extension, dominance, and influence of the biomedical model on research and practice in the aging field. While Clarke et al.'s (2003) work focuses on describing broad trends in medicine and health care, Estes and Binney (1989) looked more towards the problems that biomedicalization brings when it is applied to normal life stages," for example, aging.

A third way of conceptualizing biomedicalization is to define it not as a separate concept but rather as an extension of medicalization itself given recent changes within medical frameworks. Conrad

(2004) takes this position by pointing out that medicalization has shifted from the 1970s and 1980s in which concerns were more focused on the domination of the medical profession, social movements and interest groups, and the organization of professional activities. With the increasing attention placed on biotechnology, consumer groups, and managed care organizations since the latter part of the 1980s, it is argued that the "engines of medicalization" have shifted from physicians as the central focus to one which is more reflective of commercial and market interests (Conrad, 2004). This shift has included the proliferation of biotechnology, primarily in the form of pharmaceutical products, that has pushed medical or behavioral difficulties into the realm of medical diseases in order to justify such treatments. Increasingly off-label uses of medications have been utilized to either extend the age of treatment, for example, ADHD medications used by adults, or to identify other problem areas that can be termed diseased and in need medications. Second, consumers have played a major role in wanting and sometimes demanding medical services by self-medicalizing: that is, determining what their problems are and requesting a medical solution. Finally, Conrad (2004) points to managed care as a third force in medicalization. Because the goal of managed care is to contain costs yet provide good medical care, psychiatric medications may be encouraged as opposed to longer, more expensive therapy sessions. However, managed care organizations may not cover specific procedures that they deem are experimental or unproven. Thus, newer biotechnological approaches to medical problems may be less used, at least initially, in managed care health delivery systems. Conrad (2004) maintains that these three processes represent shifts of medicalization as opposed to a transformation of how health and health services are conceptualized. Further, he critiques Clarke et al.'s (2003) "broad-brush" approach by arguing that "...it loses sight of medicalization itself" (p. 3).

Clearly, the process of biomedicalization can be seen from a variety of perspectives. No doubt as medical sociologists and other social scientists continue to assess this process as newer biotechnical and pharmacological discoveries are made to treat an ever increasing array of diseases, the issue of whether biomedicalization is an extension of medicalization or in fact, is a unique, newer process will be determined. Regardless of how it is viewed, all perspectives recognize the ways in which biomedicalization is affecting how disease is defined and thus, what needs to be invoked to treat such conditions. Less obvious, perhaps, is the recognition that biomedicalization represents a narrower perspective than medicalization with regard to other levels of inquiry, for example, environmental, that need to be considered whenever diseases or "bad behaviors" are studied.

The New Genetics

A discussion of biomedicalization would be incomplete without acknowledging the influence of the "New Genetics" on what types of research are currently encouraged and rewarded. While it goes beyond the scope of this book to provide a detailed account of genetic research from both an historical and contemporary perspective, it is important to provide a context for us to understand better how this type of research has recently gained much prominence and has further moved the focus of the biomedicalized perspective into the realm of individual pathology.

Historically, the field of genetics can be dated back to Mendel and his famous experiments with pea plants conducted in the late 1850s. However, Francis Galton, is generally credited for taking Mendal's rules of inheritance to the level of humans and used the term "eugenics" in 1883 to describe the process of bettering the human race by controlling who should breed and under what circumstances (Allen, 1999; Conrad & Gabe, 1999). Linked to social reform and deemed scientific, the eugenics movement viewed its mission as progressive; thus, it was believed that by strongly discouraging or preventing specific individuals from breeding, who had what was considered defective or unfit genetic background, or by strongly encouraging those who possessed what was considered to be a positive genetic makeup to multiply, many social problems could be eliminated (Davenport, Laughlin, Weeks, Johnstone & Goddard, 1911). Associated with biology, eugenicists felt that the application of their breeding principles to human beings was both a scientific and moral calling to enhance the quality of the population for everyone. Thus, restricting immigration of "inferior" populations was a natural outgrowth of this movement as well as the involuntary sterilization of the mentally ill. The rise of the eugenics movement in the U.S. also changed the ways in which citizenship was viewed in relation to intellectual disability (often termed "feeblemindedness") and rights (Carey, 2003). The way in which citizenship was defined in terms of mental capacity was critical during the eugenics movement; yet, studies of this time period do not generally use citizenship theory as their framework. It is generally agreed that this focus on "bad genes" as the root of our social ills did little to alleviate the issues of poverty, criminal behavior and alcoholism during that time (Allen, 1999). While it is clear that the popularity of the eugenics movement can partially be explained by social structure and class differences, it is not surprising that funding for many of these efforts was by wealthy foundations and individuals with strong vested interests in preserving the status quo. However, its simplistic genetic explanations for social problems combined with the racial and ethnic biases displayed by the Nazi regime during World War II led to its decline. Yet, the belief that genetic properties within the individual inevitably lead to both positive and negative behaviors has not entirely disappeared.

Perhaps partly as a reaction to the eugenics movement, human behavior was often viewed as a consequence of both psychological and sociological processes separate from genetic issues following World War II and up to the 1970s. However, as molecular biology began to develop as a serious discipline, "...genetic determinism was in vogue once more" (Cunningham-Burley & Kerr, 1999, p. 651). In the 1980s, the Human Genome Project (HGP) began and promised the ability to create a 'book of man' that could potentially end genetic disease and thus enhance the quality and quantity of human life.

Martin (1999) discusses this time period in terms of the changes experienced in the field of molecular biology following the publication of a report entitled *Splicing Life* written by the Commission for the Study of Ethical Problems in Medicine, and Biomedical and Behavioral Research in 1982. This report divided genetics research into two areas: germ line therapy (deemed unethical) in which the goal is to affect future generations, and somatic gene therapy (deemed acceptable) that focuses on treatment of patients with specific illnesses. Other developments in this area include the development of the biopharmaceutical industry (in close alliance with academia) and the movement away from studying specific, classic genetic diseases to acquired disorders with a primary focus on cancer (Martin, 1999). As work in this area continued, the genetic role in cancer was becoming more prominent particularly in terms of how genes might be able to control cancerous cell growth that could potentially be cured by gene transfer. This could only be possible as the field moved more towards molecular genetics, and hence a molecular pathologic perspective, in which gene therapy including cell transplants and gene transfer were the primary treatments. Eventually, this perspective was widened to include acquired diseases such as cystic fibrosis, AIDS, arthritis and heart disease (Martin & Thomas, 1996).

The impact of these developments along with the mapping of the human genome has major implications for research and for the public. Support for genetic explanations of diseases as well as behavioral and medical conditions is evidenced by the enthusiasm for the Human Genome Project (HGP) by both the research community as well as the public at large. This signals a strong belief in the power of science to cure disease, prolong life, and enhance the quality of life. With these beliefs comes hope for not only alleviating existing problems but also preventing them in the future. Phrases have emerged that signify this hope associated with the HGP including comparisons to the "holy grail" and declaring that 'our fate is no longer in the stars but in our genes.' Thus, in addition to hopefulness, there is an almost religious fervor over what the future may hold when the mystery of genes further unfolds (Allen, 1999).

Yet, despite this excitement and enthusiasm, several warnings have been issued that point to the potential downsides of the promise of genetic research. For example, it has been argued that genes have now not only become cultural icons, but they have also become commodities in and of themselves (Nelkin & Lindee, 1995). Thus, issues of property rights, privacy, monopoly rights and commercial development rights come into play (Everett, 2003). Moreover, the social meaning of gene therapy, its potential, and how this may affect how disabled people are valued or not valued in our society is another issue (Shakespeare, 1999). Allen (1999) argues that the promotion of genetics as explanations for diseases and behaviors is more the product of economic and social contexts than compelling scientific findings.

The expanding involvement of the private sector in terms of biotech startup companies in the development of the technology to study molecular genetics, the profit motives of these companies, and their intimate association with academic researchers sets the context for ethical concerns that arise in this arena, particularly conflicts of interests (Krimsky, 2003). Increasing concern has been expressed about genetic determinism, and its potential for discrimination in the realm of employment, health insurance coverage and life insurance coverage, and whether genetic data should be readily available to outside sources (Nelkin & Andrews, 1999; Nelkin & Tancredi, 1994). Expansion of the medical sphere to include "bad" behaviors as ill-nesses has been discussed thoroughly in the medicalization litera-ture (e.g., Conrad & Schneider, 1992). Expansion of genetic research to include "bad" behaviors takes this approach one step further in terms of limiting illness or disease to microbiologic mechanisms clearly located within the individual with very little credence given to the importance of the environment. For example, by bio-medicalizing Attention Deficit Disorder (ADHD) so that children diagnosed with this pathology are likely to be treated with psycho-tropic drugs such as Ritalin, other external factors that may contrib-ute to the problem, such as inadequate teachers, lack of quality time with parents, poor nutrition are disregarded (Allen, 1999; Conrad, 1975; Singh, 2002). The success of this process has led to a rapid expansion of the Ritalin market to children and an expansion of the ADHD diagnosis to adults (Conrad & Potter, 2000). Other case ex-amples, discussed below, share a concern about the focus on indi-vidual pathology (now from a molecular genetics framework) and the resultant lack of attention towards social and environmental forces that help shape such behaviors. Duster (2003) provides a strong warning concerning current genetic research. He argues that while the front door to eugenics is no longer available given the events of World War II, some aspects of the New Genetics may lead one to question whether eugenics is reoccurring through the "back door." By this he means when taken in isolation, the ability to detect dis-ease through the tracing of genetic structures has several advan-tages; yet, as the government (state and federal) moves more to-wards mandatory reporting of chromosomal abnormalities, there is always the danger that this information may be used against those who are the most vulnerable based on income, ethnicity or gender. What may be lost in these discussions are the political and structural frameworks that account for many of the serious health discrepan-cies among marginalized populations.

Case Examples of Biomedicalization

The following case examples of biomedicalization are not intended to represent all fields or areas in which this process is occurring, but rather to illustrate with a few areas the extent to which bio-medicalization has been experienced. Three areas will be discussed that illustrate the wide range of areas biomedicalization affects: aging, psychiatric/mental health, and women's health research, policy and practice.

Aging. One of the first references to biomedicalization was made in an article written in 1989 by Carroll Estes and Elizabeth Binney that specifically examined the dangers and problems of this process within the field of aging (Estes & Binney, 1989). While Clarke et al. (2003) acknowledge the earlier use of this term by these researchers as well as others (Lyman, 1989; Vertinsky, 1991), they appear to define biomedicalization by its association on technoscience. To some extent, this is a semantic point in that within the aging field, the focus is moving towards individual organic pathology and interventions, much of which could not be diagnosed or treated without the extensive input of technoscientific tests and treatments. Thus, for the purposes of this discussion, biomedicalization will encompass the larger, structural technoscientific structures discussed by Clarke et al. (2003) as well as the process by which these forces have played a role in changing on specific areas are defined and conceptualized.

Perhaps the most striking feature of the Estes and Binney (1989) article is the shift and reformulation of aging from a natural lifecourse event to a pathologic condition. Estes & Binney (1989) use the term "methodological individualism," defined as "...as a strategy of theory construction and investigation that seeks to explain any social institution or phenomenon using individuals as the unit of analysis" (p 588), to further demonstrate the detrimental effects of a scientific reductionism which excludes broader social and environmental issues. They define biomedicalization in terms of not only the medicalizing of aging, but also the dominance of the biomedical model that equates the aging process with an illness, and thus as a lifestage, it has become abnormal. Moreover, Estes and Binney (1989) argue strongly that this process has had a profound influence not only on how aging is handled by our society, but also over "...other research, policymaking, and the way we think about aging and even science, as it is defined and evaluated in terms of a biomedical structure of thought" (p 588). Unique to their article is an in-depth discussion of the effects of the biomedicalization of aging on aging policy, research, funding, and training on aging, and the lay public's perception of aging. Their work, supported other research at the time (Kerin, Estes, & Douglass, 1989), indicates that biomedical education and research gets a larger proportion of support than social and behavioral efforts. Federal research funding "...favour[s] medically defined problems, approached from the perspective of the scientific concerns of biology, chemistry, immunology, and other "hard" sciences" (Estes and Binney, 1989; p 591). This has meant that social scientists that conduct research on aging have increasingly had to adopt a medical persona to legitimize its existence and perhaps make it more comfortable for funders to approve their research.

While no one argues that the biomedical model should be eliminated from research on aging, its dominance to the exclusion of behavioral and social approaches is problematic. In her work on senile dementia, Lyman (1989) points to how this dominance of the biomedical model has led to a minimization of social factors that have a significant impact on treatment, caregiver relationships, and subsequently on the social and biological course of dementia for individuals. The biomedicalization of aging with its resultant focus on Alzheimer's disease, purported to be in epidemic proportions among the elderly, has also led to what one author describes as an "apocalyptic demography" perpetrated by political concerns as opposed to sound epidemiologic data (Robertson, 1990). Royall (2003) refers to this process as "the "Alzheimerization" of dementia research" as reflected in the title of his article.

The problem created by the dominance of the biomedical model in the aging field has been termed a crisis by Estes and Binney (1989) for two reasons. Biomedical interventions may have contributed to the extension of the lifespan but appear to have done little to enhance the quality of these added years. Thus, social issues of quality of life and social costs are not included within this perspective. Second, we need to ask if the lack of attention by the biomedical model to larger structural issues itself contributes to illness. The focus has primarily been on individual levels of prevention to the exclusion of larger political considerations. The consequences of such exclusions may be important both for geriatric patients and for social gerontology.

Psychiatry/Mental Health. It can be argued that concerns with mental health, and more particularly the diagnosis and treatment of mental illness, is, to a large extent, a positive outcome of the historical process of medicalization that has been discussed earlier. What was not that long ago viewed as "being possessed," or "irrational," "mad," or "crazy" has evolved from the frightening and misunderstood to an illness (or set of illnesses) which, with varying degrees of success, is at least partially understandable and potentially treatable even though it is often stigmatized (Foucault, 1965). Mental illness, at least in the Western tradition, came to be seen explained by theology, was thought of less as criminal, and more as a series of clinical entities, often still of unknown etiology, which could and should be treated to provide less discomfort for its sufferers who became patients. Thus, persons with mental illness became less a matter of possession or madness and more a matter of someone who could be restored to some semblance of psychosocial well being with appropriate counseling and therapy (Solomon, 2001). Yet, with a lack of evidence-based practice modalities to legitimately claim that mental illness treatments were successful, a high degree of skepticism prevailed from within and outside the mental health professions.

The coming of "brain science" overturned some of this scepticism. The "mysteries of the mind" began to yield to clinical neuroscience, seeking to meld what we knew from psychology with the rapidly increasing discoveries and knowledge coming from behavioral genetics, biochemistry, neuroanatomy, psychopharmacology, studies of cognition and emotion in the brain, and, biology (Brown, 1996). With excitement, claims have been made for many truly wondrous advances; soon, it was felt, we would have solid genetic explanations, with treatments not far behind, for various forms of mental illness including episodic and major depression, schizophrenia, and other "chronic mental illness." As a result of these enthusiasms and as a result of seeking to reduce stigma, "chronic mental illness" was termed "severe and persistent mental illness" and that terminology better represents the promises of successful treatment based in a biomedical approach. What was previously referred to as "mysteries of the mind" more and more became "brain science," and, with it, a plethora of both new drugs, and thus new treatments for mental illness (Kandel, 1993)

Cohen (1993) terms the growth of the biomedical model in psychiatry as biological reductionism, defined as "explanations of phenomena occurring at several levels (e.g., social, psychological) that are sought at a single level (biology)" (p 510). He goes on to strongly argue that the result of this process in psychiatry has been a movement towards individualistic causes and treatments of psychiatric problems and a de-emphasis on the social environment. Beyond the myth that biological psychiatry is objective, Cohen (1993) points out that underlying biological reductionism is the idea that treatment and causality are inextricably linked. That is, it is assumed that because a particular drug intervention may lessen or completely end specific symptomatology or condition, then the underlying problem is primarily biological by definition. This "faulty logic" clearly minimizes the role of environment in mental illness and once again focuses on the individual almost exclusively as the primary focus of intervention.

Many other researchers have been critical of the medical and biomedical models of psychiatry e.g., (Double, 2003; Holmes, 2001; Kutchins & Kirk, 1997; Moncrieff, 1997) with perhaps the most famous being Szasz (1961; 1970) who, several decades ago, wrote about the social construction of mental illness and questioned its very existence. Most recently in the UK, a network of psychiatrists created The Critical Psychiatry Network (www.criticalpsychiatry.co.uk). This network

"...challenges the dominance of clinical neuroscience in psychiatry (but does not exclude it); it introduces a strong ethical perspective on psychiatric knowledge and practice; it politicizes mental health issues. Critical psychiatry is deeply skeptical about the reductionist claims of neuroscience to explain psychosis and other forms of emotional distress. It follows that we are skeptical about the claims of the pharmaceutical industry for the role psychotropic drugs in the 'treatment' of psychiatric conditions. Like other psychiatrists we use drugs, but we see them as having a minor role in the resolution of psychosis or depression. We attach greater importance to dealing with social factors, such as unemployment, bad housing, poverty, stigma and social isolation. Most people who use psychiatric services regard these factors as more important than drugs. We reject the medical model in psychiatry and prefer a social model, which we find more appropriate in a multi-cultural society characterized by deep inequalities."

Despite this effort to reject the biomedical model, others have noted how biomedical approaches, particularly genetics, have been presented in the media as critical and groundbreaking in the understanding mental illness, particularly bipolar disorders and schizophrenia (Conrad, 2001). Increasingly, news stories carry the message that mental illness can be explained by genetics, that specific genes linked to specific mental illnesses will be found, and that once they are found, only good outcomes such as recovery and prevention can be achieved. Conrad (2001) terms this way of thinking about mental illness as "genetic optimism" which appears to be quite resilient even though these claims cannot be substantiated by replicated research. This equation of mental illness with biochemical or genetic underpinnings is very apparent in the area of schizophrenia in which the discourse is increasingly biomedical in close alliance with science and therefore "progress" (Boyle, 2002). It is thus no surprise that the central mode of training new psychiatrists at major research and teaching programs includes learning how to rapidly diagnose patients paired with rapid prescription of medication (Luhrmann, 2000)

*Women's health.*Multiple phases in women's lives have been placed in the realm of medicine and biotechnology ranging from pregnancy, childbirth, premenstrual and perimenopausal symptoms, and menopause (Purdy, 2001). The line between what conditions may or may not need medical intervention is often blurred. More recently, additional areas are being addressed under women's health from the perspective of the biomedical model, such as sexual dysfunction, infertility, and breast cancer risk. With these newer areas, there is a greater recognition of the biotechnical developments that have contributed to addressing these issues as pathological. To provide a context for the medicalization of women's life stages, menopause will be briefly discussed followed by more recent work in the area of women's sexual dysfunction.

Menopause, as a natural life stage for women, should be a fairly neutral concept. Yet, with the rise of hormone replacement therapy beginning in the 1960s, this natural lifestage came to be viewed as a disease for which medications are the optimal treatment. It is interesting to note Beyene and others have found that typical Western symptoms of menopause, such as hot flashes, are not necessarily universal despite similar estrogen levels in menopausal women, and, when symptoms of menopause are found across cultures, they are may be viewed in a more positive light (Beyene, 1986; Martin, Block, Sanchez, Arnaud, & Beyene, 1993). When menopause is addressed as pathology caused by physiological mechanisms that result in discomfort that "needs" to be treated with pharmacological interventions, and then clearly the dominant framework is biomedical. Often, cultural differences among groups are absent from the discussion. Discussed in terms of a 'pharmaceutical' discourse, Coupland and Williams (2002) found that this is only one of at least three discourses on menopause ('alternative' therapy discourse and emancipatory feminist discourse); yet, it appears to be the more dominant one with its own distinct ideology.

In looking at how menopause is conceptualized by analyzing publication trends in both medical and psychological journals, Rostosky and Travis (1996) noted the dominance of the biomedical model to explain, diagnose and treat this normal lifestage. They argue that this focus on the biomedical aspects of menopause supports the view that women deviate from the normal, e.g., the "opposite" sex, the "sicker" sex, the "weaker" sex, in that their hormone changes vary significantly from what is experienced by the "normal", e.g., men. By pathologizing menopause, women are seen as being disadvantaged by their physiology and are thus in need of some kind of intervention (Rostosky & Travis, 1996). It is not surprising that menopause, following from the medical literature, has also been framed negatively in the popular media. During 1981 through 1994 in articles published in *The Readers Digest*, Gannon and Stevens (1998) identified 350 instances of negative experiences of menopause as compared to only 27 positive experiences in popular articles in the media. These results would have probably been quite different had menopause been studied 50 years ago.

Female sexual dysfunction is a relatively new area of women's health that has biomedicalized perhaps following from the trend in male sexuality with the introduction of Viagara (Bancroft, 2002). With estimates of sexual dysfunction of 43 percent of women and 31 percent of men (Laumann, Paik, & Rosen, 1999), this is an area that is "ripe" for medical intervention. Bancroft (2002) makes a convincing argument that during the last century, female sexuality, for the most part, has been ignored as a specific problem and when discussed, it is seen as a side effect of invasive medical procedures such as hysterectomy. This has recently changed perhaps as a consequence of the development of new drugs that may positively affect female sexual dysfunction. Bancroft (2002) cautions that there is a fine line between sexual problems and sexual dysfunction. By focusing on dysfunction (a medical diagnosis) and treating the condition with medical interventions, larger issues pertaining to interpersonal relationships and societal expectations of women's sexuality are at best, minimized or at worst, totally ignored (Moynihan, 2003).

In a study of how Viagara, a drug approved by the Food and Drug Administration (FDA) in 1998 for the treatment of erectile dysfunction, and Eros, a clitoral therapy device approved in 2000 by the FDA, were marketed and promoted by the pharmaceutical industry, Fishman and Mamo (2002) illustrate the powerful effects of biomedicalization on images of female and male sexuality. These effects include the "diseasing" of female sexual dysfunction, the promotion of sexuality based on assumptions around heterosexuality and coupled sexual intercourse (between a man and a woman only) as the "gold standard"(Fishman & Mamo, 2002). Pointing out that Viagara is just another example of how biomedical approaches have increasingly become part of our everyday existence, biomedical solutions to issues such as sexual dysfunction can be seen as gendered technologies which support the dominant, status quo version of how and with whom sexual relations should occur (Fishman & Mamo, 2002; Mamo & Fishman, 2001).

Summary

The purpose of this chapter was threefold. First, it focused on the rise of medicalization since the 1970s and the subsequent rise of biomedicalization over the last two decades. Second, it examined how biomedicalization can be conceptualized as a significant extension of medicalization that looks not only at how the mechanism of disease has become exceedingly pervasive, but also addresses how biomedicalization increasingly focuses on the individual as the unit of analysis for the diagnosis and treatment and even for prevention thereby moving away from larger, environmental explanations. Finally, examples of biomedicalization were presented that illustrate its recognition appeal in multiple fields.

That biomedicalization is occurring is not disputed within or across disciplines. Yet the lingering question is what this biomedicalization may mean for how we approach current and future behaviors that may be deemed "diseases" by pressure brought upon the medical profession through intense marketing of biomedical solutions. Recent work in the area of pharmacological interventions has been critical of industry-funded studies of new medications and questions the reliability of data when how the study is design, how the subjects are selection, how the overall research procedure are conducted and how the results are interpreted and disseminatd is greatly influenced by the pharmaceutical industry (Lemmens, 2004). Moreover, it has been argued marketing of medications directly to physicians by the pharmaceutical industry and the sponsoring of 60 percent of educational programs that often provides Continuing Medical Education credits it the primary way that physicians are educated about which drugs to prescribe to their patients (Elliott, 2004). Clearly, the agenda of the pharmaceutical industry plays a major role in how diseases are defined and medications are marketed for substantial profits.

The remaining chapters in this book will focus on the alcohol field in the U.S. as a case study of biomedicalization. While clearly there are several areas that could have been used as case studies, the alcohol field is somewhat unique in how alcoholism has shifted from "bad behavior" to a disease with less participation by physicians and more involvement by self-help groups, e.g., Alcoholics Anonymous. The next chapter will provide the historical context for the medicalization of the alcohol field with a history of how and why "addiction" was discovered, the impact of the disease model in the alcohol field, the development and growth of Alcoholics Anonymous, and its expansion to a wide range of people need of help. This chapter will lay the groundwork for cultural and institutional changes that have occurred in the last 15 years that have moved the field to a more central position in the biomedical field.

References

Allen, G. E. (1999). Genetics, eugenics and the medicalization of social behavior. *Endeavour, 23*(2), 10-19.

AMA. (1847). *Code of Medical Ethics of the American Medical Association.* http://www.ama-assn.org/ama/upload/mm/369/1847code.pdf: Originally Adopted at the Adjourned Meeting of the National Medical Convention in Philadelphia.

Armstrong, M. A., Midanik, L. T., Klatsky, A. L., & Lazere, A. (2001). Utilization of Health Services Among Patients Referred to an Alcohol Treatment Program. *Substance Use and Misuse, 36*, 1781-1793.

Bancroft, J. (2002). The medicalization of female sexual dysfunction: The need for caution. *Archives of Sexual Behavior, 31*(5), 451-455.

Beyene, Y. (1986). Cultural significance and physiological manifestations of menopause. A biocultural analysis. *Cult Med Psychiatry, 10*(1), 47-71.

Boyle, M. (2002). *Schizophrenia: A Scientific Delusion?* New York: Routledge.

Brown, E. R. (1979). *Rockefeller Medicine Man: Capitalism and Medical Care in America.* Berkeley, CA: UC Press.

Brown, G. W. (1996). Genetics of Depression: A Social Science Perspective. *International Review of Psychiatry, 8*, 387-401.

Carey, A. C. (2003). Beyond the medical model: A reconsideration of 'feeblemindedness,' citizenship, and eugenic restrictions. *Disability & Society, 18*(4), 411-430.

Clarke, A., Shim, J. K., Mamo, L., Fosket, J. R., & Fishman, J. R. (2003). Biomedicalization: Technoscientific transformations of health, illness, and U.S. biomedicine. *American Sociological Review, 68*, 161-194.

Cohen, C. I. (1993). The biomedicalization of psychiatry: A critical overview. *Community Mental Health Journal, 29*, 509-521.

Conrad, P. (1975). The discovery of hyperkinesis: Notes on the medicalization of deviant behavior. *Social Problems, 23*, 12-21.

Conrad, P. (1992). Medicalization and social control. *Annual Review of Sociology, 18*, 209-232.

Conrad, P. (2001). Genetic optimism: Framing genes and mental illness in the news. *Culture, Medicine and Psychiatry, 25*, 225-247.

Conrad, P. (2004). *The Shifting Engines of Medicalization*. Paper presented at the American Sociological Association, San Francisco, CA.

Conrad, P., & Gabe, J. (1999). Introduction: Sociological perspectives on the New Genetics: An overview. In *Sociological Perspectives on the New Genetics* (pp. 1-12). Malden, MA: Blackwell Publishers.

Conrad, P., & Potter, D. (2000). From hyperative chldren to ADHD adults: Observations on the expansion of medical categories. *Social Problems, 47*(4), 559-582.

Conrad, P., & Schneider, J. W. (1992). *Deviance and medicalization: From badness to sickness*. Philadephia: Temple University Press.

Conrad, P., & Schneider, J. W. (2001). Professsionalization, monopoly and the structure of medical practice. In P. Conrad (Ed.), *The Sociology of Health and ILlness. Critical Perspectives* (Sixth ed., pp. 156-162). New York: Worth Publishers.

Coupland, J., & Williams, A. (2002). Conflicting discourses, shifting ideologies: Pharmaceutical, 'alternative' and feminist emancipatory texts on the menopause. *Discourse & Society, 13*(4), 419-445.

Cunningham-Burley, S., & Kerr, A. (1999). Defining the 'social:' Towards an understanding of scientific and medical discourses on the social aspects of the new human genetics. *Sociology of Health & Illness, 21*(5), 647-668.

Davenport, C. B., Laughlin, H. H., Weeks, D. F., Johnstone, E. R., & Goddard, H. H. (1911). *The Study of Human Heredity*. New York: Cold Spring Harbor.

Donahoe, T., & Johnson, N. (1986). *Foul Play: Drug Abuse in Sports*. Oxford: Blackwell Scientific Publications.

Double, D. B. (2003). Can a biomedical approach to psychiatric practice be justified? *Journal of Child and Family Studies, 12*(4), 379-384.

Duster, T. (2003). *Backdoor to Eugenics* (Second ed.). New York: Routledge.

Elliott, C. (2004). Pharma goes to the laundry: Public relations and the business of medical education. *Hastings Center Report, 34*(5), 18-23.

Engel, G. (1977). The need for a new medical model: A challenge for biomedicine. *Science, 196*, 129-136.

Engel, G. (1980). The clinical application of the biopsychosocial model. *American Journal of Psychiatry, 137*(5), 535-544.

Estes, C. L., & Binney, E. A. (1989). The biomedicalization of aging: Dangers and dilemmas. *Gerontologist, 29*, 587-596.

Eugene Traynor, Petitioner v. Thomas K. Turnage, Administrator, Veterans' Administation and The Veterans' Administration, Supreme Court 2631-2638 (1988).

Everett, M. (2003). The social life of genes: Privacy, property and the new genetics. *Social Science & Medicine, 56*, 53-65.

Fingarette, H. (1988). *Heavy drinking: the myth of alcoholism as a disease*. Berkeley: University of California Press.

Fishman, J. R., & Mamo, L. (2002). What's in a disorder? A cultural analysis of the medical and pharmaceutical constructions of male and female sexual dysfunction. *Women and Therapy, 24*, 179-193.

Flexner, A. (1910). *Medical Education in the United States and Canada. A Report to The Carnegie Foundation for the Advancement of Teaching* (No. Bulletin Number Four). New York: The Carnegie Foundation for the Advancement of Teaching.

Foucault, M. (1965). *Madness and Civilization*. New York: Vintage.

Friedson, E. (1970). *Profession of Medicine*. New York: Dodd, Mean, & Co.

Gambrill, E. (in press). Critical Thinking, Evidenced-Based Practice, and Mental Health. In S. Kirk (Ed.), *Mental disorders in the social environment: Critical perspectives*. New York: Columbia.

Gannon, L., & Stevens, J. (1998). Portraits of menopause in the mass media. *Women Health, 27*(3), 1-15.

Hawkins, A. H. (1994). Reforming the biomedical model: Finding a successor model or going beyond the paradigms? *Advances, 10*, 55-56.

Hewa, S., & Hetherington, R. W. (1995). Specialists without spirit: Limitations of the mechanistic biomedical model. *Theoretical Medicine, 16*, 129-139.

Holmes, C. A. (2001). Guest Editorial. *Journal of Psychiatric and Mental Health Nursing, 8*, 379-381.

Illich, I. (1975). *Medical nemesis : the expropriation of health*. New York: Pantheon Books.

James P. McKelvey, Petitioner v. Thomas K. Turnage, Administrator of Veterans' Affairs, et al., Supreme Court 2631-2638 (1988).

Kandel, E. (1993). A new intellectual framework for psychiatry. *American Journal of Psychiatry, 155*, 457-469.

Kerin, P. B., Estes, C. L., & Douglass, E. B. (1989). Federal funding for aging education and research: a decade analysis. *The Gerontologist, 29*(5), 606-614.

Kirk, S. A., & Kutchins, H. (1992). *The Selling of DSM: The Rhetoric of Science in Psychiatry*. New York: Aldine de Gruyter.

Krimsky, S. (2003). *Science in the private interest: Has the lure of profits corrupted biomedical research*. Lanham, MD: Rowman and Littlefield.

Kutchins, H., & Kirk, S. (1997). *Making Us Crazy: DSM: The Psychiatric Bible and the Creation of Mental Disorders*. New York: Free Press.

Laumann, E. O., Paik, A., & Rosen, R. C. (1999). Sexual dysfunction in the United States: Prevalence and predictors. *JAMA, 281*(6), 537-544.

Lemmens, T. (2004). Piercing the veil of corporate secrecy about clinical trials. *Hastings Center Report, 34*(5), 15-18.

Luhrmann, T. M. (2000). *Of Two Minds: The Growing Disorder in American Psychiatry*. New York: Knopf.

Lyman, K. A. (1989). Bringing the social back in: A critique of the biomedicalization of dementia. *Gerontologist, 29*, 597-605.

Mamo, L., & Fishman, J. R. (2001). Potency in all the right places: Viagara as a technology of the gendered body. *Body and Society, 7*, 13-35.

Martin, M. C., Block, J. E., Sanchez, S. D., Arnaud, C. D., & Beyene, Y. (1993). Menopause without symptoms: the endocrinology of menopause among rural Mayan Indians. *Am J Obstet Gynecol, 168*, 1839-1843.

Martin, P. (1999). Genes as drugs: The social shaping of gene therapy and the reconstruction of genetic disease. In P. Conrad & J. Gabe (Eds.), *Sociological Perspectives on the New Genetics* (pp. 15-35). Malden: MA: Blackwell Publishers.

Martin, P., & Thomas, S. M. (1996). *The Development of Gene Therapy in Europe and the United States: A Comparative Analysis* (No. STEEP Special Report No. 5). Brighton: Science Policy Research Unit, University of Sussex.

McKinlay, J. B., & Stoeckle, J. D. (2001). Corporatization and the Social Transformation of Doctoring. In P. Conrad (Ed.), *The Sociology of Health and Illness* (Sixth ed., pp. 175-186). New York: Worth Publishers.

Moncrieff, J. (1997). Psychiatric imperialism: the medicalisation of modern living. *Soundings, 6*.

Moynihan, R. (2003). The making of a disease: female sexual dysfunction. *British Medical Journal, 326*, 45-47.

Moynihan, R., Heath, I., & Henry, D. (2002). Selling sickness: the pharmaceutical industry and disease mongering. *British Medical Journal, 324*(886-891).

Nelkin, D., & Andrews, L. (1999). DNA identification and surveillance creep. In P. Conrad & J. Gabe (Eds.), *Sociological Perspectives on the New Genetics* (pp. 191-208). Malden, MA: Blackwell Publishers.

Nelkin, D., & Lindee, M. S. (1995). *The DNA Mystique: The Gene as a Cultural Icon*. New York: W.H. Freemen.

Nelkin, D., & Tancredi, L. (1994). *Dangerous diagnostics: the social power of biological information.* Chicago: University of Chicago Press.

Parsons, T. (1975). The sick role and the role of the physician reconsidered. *Milbank Memorial Fund Quarterly, summer*, 257-278.

Purdy, L. (2001). Medicalization, medical necessity, and feminist medicine. *Bioethics, 15*(3), 248-261.

Robertson, A. (1990). The politics of Alzheimer's disease: A case study in apocalyptic demography. *International Journal of Health Services, 20*(3), 429-442.

Rostosky, S. S., & Travis, C. B. (1996). Menopause research and the dominance of the biomedical model, 1984-1994. *Psychology of Women Quarterly, 20*, 285-312.

Royall, D. (2003). The "Alzheimerization" of dimentia research. *Journal of the American Geriatrics Association, 51*(2), 277-281.

Shakespeare, T. (1999). Losing the plot? medical and activist discourses of the contemporary genetics and disability. In P. Conrad & J. Gabe (Eds.), *Sociological Perspectives on the New Genetics.* Malden, MA: Blackwell Publishers.

Sinclair, U. (1906). *The Jungle.* New York: Airmont.

Singh, I. (2002). Bad boys, good mothers, and the "miracle" of Ritalin. *Science in Context, 15*(4), 577-603.

Smith, R. (2002). In search of "non-disease". *British Medical Journal, 324*(13 April), 883-885.

Solomon, A. (2001). *The Noonday Demon: An Atlas of Depression.* New York: Scribner.

Starr, P. (1982). *The Social Transformation of American Medicine.* New York: Basic Books.

Stevens, R. (1971). *American Medicine and the Public Interest.* New Haven, CT: Yale University Press.

Szasz, T. S. (1961). *The Myth of Mental Illness; Foundations of a Theory of Personal Conduct.* New York: Hoeber-Harper.

Szasz, T. S. (1970). *The Manufacture of Madness; A Comparative Study of the Inquisition and the Mental Health Movement.* New York: Harper & Row.

Taylor, J., Stuart. (1988, April 21). Alcoholics Lose Suite Over Some Veterans Benefits. *The New York Times.*

Tesh, S. N. (1988). *Hidden arguments. Political ideology and disease prevention policy.* New Brunswick, NJ: Rutgers University Press.

Verroken, M. (1996). Drug use and abuse in sport. In D. R. Mottram (Ed.), *Drugs in Sport* (Second ed.). London: E.& F. N. Spon.

Vertinsky, P. (1991). Old age, gender and physical activity: The biomedicalization of aging. *Journal of Sport History, 18*(1), 64-80.

Waddington, I. (2000). *Sport, Health and Drugs.* New York: E & FN Spon.

Zola, I. K. (1972). Medicine as an institution of social control. *Sociological Review, 20*(4), 487-504.

Zola, I. K. (1975). In the name of health and illness: On some socio-political consequences of medical influence. *Social Science & Medicine, 9*(2), 83-87.

3

Precursors of Biomedicalization in Alcohol Studies: Medicalization and the Disease Model of Alcoholism

...the rise for the disease concept of alcoholism illustrates that the process of collective medical definition need not necessarily be sustained primarily by medical personnel. The historical development of medical definitions of deviant drinking involves powerful nonmedical groups, individuals, and organizations whose moral, political, status, and/or professional interests were served by such definitional change. Although physicians were not hapless bystanders in this process, they were not the leading crusaders. (Conrad & Schneider, 1992, p 73)

Introduction

The last chapter delineated the general rise of medicalization and the more recent development of biomedicalization in the health field and briefly included some case examples of this process. This chapter will focus in more depth on one specific area, alcoholism, and will assess the development and assumptions underlying two very intertwined movements that have greatly influenced the substance abuse field: first, the medicalization of deviance and second, the growth of the disease model of alcoholism. The disease model has had a profound influence in the alcohol field and subsequently on the type of research that has been encouraged and conducted. The disease model has greatly influenced the treatment arena; yet little has been written about how its assumptions have provided a very specific context for the types of alcohol research projects conducted. How one defines or labels an issue (ranging from who is an alcoholic to who is most likely to report an alcohol-related problem) is based on the assumptions one makes about the very people who are believed to manifest this problem. To understand better the interplay between alcohol research and the social and political context within which it occurs, a brief discussion of the history and development of the disease model is presented. By examining critically the presuppositions behind the disease model of alcoholism and the groups that have adopted this model, we can appreciate why social scientists working in the alcohol field who do not necessarily adopt these assumptions may have more difficulty in obtaining funding for their research.

One definitional point needs to be made in terms of the terminology used to describe the disease model as compared to other models of alcoholism. In the alcohol literature, distinctions have been made between medical models and disease models. Siegler, Osmond and Newell (1968) argue that there are two different medical models (the "old" and the "new" medical models) that co-exist with the disease model; both of these medical models share some commonalities with the disease model but also have distinct differences. In both medical models, alcoholism is defined as a disease that is progressive and often fatal. However, the "old" medical model assumes a moral etiology; that is, "alcoholics seem to be unable to control themselves" (p. 580) whereas the "new" medical model assumes a disease mechanism (body chemistry, defect in metabolism, etc.) that leads to addiction to alcohol. The two medical models differ greatly in their approach to treatment. The "old" medical model argues for a treatment plan that is aimed at bringing the alcoholic to a place when he or she may engage in social drinking; for the "new" medical model, abstention is the goal of treatment. Finally, in the "old" medical model, physicians assume the primary role of providers of care initially (with the family) to attempt to get the alcoholic to drink socially or, if this is not possible, to abstain from alcohol entirely. However, if this treatment route is not successful, physicians can absolve themselves and transfer their responsibilities to other non-medical providers or groups such as the clergy or A.A. In the "new" medical model, the physician maintains the provider role in conjunction with other personnel such as nurses, psychotherapists, social workers. The Alcoholics Anonymous Model as described by Siegler, Osmond, and Newell (1968) also views alcoholism as a disease, yet views AA (and, by definition, abstinence) as the only sound treatment option. From the perspective of AA, "Medical doctors should treat alcoholics, but they often give damaging advice and unsound treatment, due to ignorance about alcoholism" (Siegler, Osmond & Newell, 1968, p. 577).

Although there have been other articles that have explicated different models of alcoholism (Goode, 1999; Keller, 1990), there is general agreement that alcoholism is a disease that may or may not include medical intervention. While a more detailed discussion of the development of the disease model of alcoholism is provided later on in this chapter, it is important to note that "disease" is an amorphous term. Some regard the disease model to be exclusively the purview of Alcoholics Anonymous. Others regard the disease model as more integrative and inclusive of other models. Still others argue that whenever abstinence is the goal for alcoholism treatment, the disease model is invoked. In this book, "the disease model of alcoholism" will be expansive in the sense that it will refer to a generalized model that has the following specific features: it is individualistic, loss of control is its defining factor, it is a constellation of symptoms that are diagnosable, and abstinence is the primary goal of treatment. While this defines the Alcoholics Anonymous model, some or all of these precepts are also used in conjunction with other treatment modalities such as cognitive behavioral models.

The first part of this chapter will focus on the medicalization of alcoholism starting with the history of the notion of addiction and moving towards the modern alcoholism movement beginning in the 1940s. Special attention will be given to the "empirical" development of the disease model by Jellinek and the assumptions behind the measures used to create the phases of addiction as he described them. The next section will be a discussion of how the disease model grew and expanded to other groups of individuals in need of help, for example, the Al-Anon, Adult Children of Alcoholics and other co-dependency movements. Specific emphasis will be placed on the particular assumptions underlying these movements and how they have served to expand their treatment markets by expanding the disease definition.

History of Addiction as a Socially Constructed Problem

There have been several excellent historical accounts that trace the social, economic, and political threads that led to the development of the concept of addiction and thus to the disease model of alcoholism. This chapter will draw heavily on these books and articles and will attempt to summarize the various ways in which addiction to alcohol became a dominant way to conceptualize excessive alcohol use.

There is general agreement that in early America during colonial times, immoderate use of alcohol was the norm. Excessive drinking to the point of drunkenness was a common and acceptable behavior (Lender & Martin, 1987), and it has been noted that "Liquor was food, medicine and social lubricant, and even such a Puritan Devine as Cotton Mather called it the 'good creature of God'" (Levine, 1978, p. 145). While heavy drinking was somewhat universal in colonial times with high consumption levels the norm, it was not labeled as a problem per se but rather as part of life with some inconveniences. Lender and Martin (1987) point out that most of this drinking was tied directly to family and community. It was the norms of this traditional, deferent society that dictated what type of behavior would be accepted as well as how the society would handle unacceptable behavior. "Most people restricted their consumption primarily to the use of beer and cider; they very rarely became problem drinkers" (p. 15). Addiction, as we currently define it, did not exist in Colonial America—excessive drinking did.

Levine (1978) argues that the construct of addiction was developed first by physicians as exemplified in Dr. Benjamin Rush's pamphlet entitled *An Inquiry into the Effects of Ardent Spirits upon the Human Body and Mind with an Account of the Means of Preventing and of the Remedies for Curing Them*, published originally in 1784. In this brief document, a strong medical case was made against consuming liquors although Rush had no arguments against moderate consumption of beer and wine which he believed was health enhancing (Grob, 1981; Rush, 1814). Most importantly, he termed chronic drunkenness (from liquor) an "odious disease" which had eleven symptoms—most of which were behavioral: (1) unusual garrulity; (2) unusual silence; (3) captiousness and a disposition to quarrel; (4) uncommon good humour, and an insipid simpering, or laugh; (5) profane swearing, and cursing; (6) a disclosure of their own, or other people's secrets; (7) a rude disposition to tell those persons in company whom they know, their faults; (8) a certain immodest actions; (9) a clipping of words; (10) fighting; (11) certain extravagant acts which indicate a temporary fit of madness. Rush's document emphatically declared that chronic drunkenness was progressive and eventually led the drinker to increasingly serious health problems and problems of the "mind."

It was no surprise that the Temperance Movement viewed Benjamin Rush as its core leader. Most importantly, his work provided the legitimate endorsement needed to appeal to the mass public. Rush's views also codified a clear eventual direction for the Movement – one that strongly argued that the evil was the alcohol itself and did not consider either the external or internal conditions of the drinker as important factors.

There were several interrelated forces in the development of the Temperance movement, which originally advocated for moderate or temperant drinking, and then moved towards total prohibition. First, by the early 1700s, there was an overall shift of beverage preference in the population from cider/beer to spirits. Lender and Martin (1987) make the argument that this transition from beer/cider to rum was symbolic of the colonials' movement from "Old World" identification to the "New World." As settlement moved further west, grain whiskeys became the favorite alcohol primarily because it was plentiful and easier to transport. As time passed, its use grew more prevalent. Second, and seen as very much related to the first point, the general feeling was that traditional social values shared by the population were being lost, and that social order was threatened. The fear of social instability was linked with increasing use of spirits and its associated, sometimes negative, behaviors. Benjamin Rush captured this view in his writings. "Ardent spirits" became the enemy and the clear cause of a variety of individual as well as social ills.

A third force contributing to the American Temperance Movement was the first large-scale wave of immigration in the first half of the nineteenth century. This new wave of immigrants, primarily Catholics from Ireland, brought with them a very different culture as well as different drinking patterns that threatened the native-born Americans. Heavier drinking for Irish men was the norm, and as anti-Irish discrimination increased, "Irish drinking took on a greater symbolic and emotional significance...the immigrants seized upon drinking as a major symbol of ethnic loyalty" (Lender & Martin, 1987, p. 60). With industrialization and the continued need for additional labor, more groups immigrated to America and also brought with them their own beverage preferences and drinking styles. Finally, the rise of "skid rows" in many cities and the urban saloon became symbols of the negative effects of excessive alcohol use. While it has been pointed out that the saloon was the center of political organizing and activism by workers (often immigrants) and thus was seen as a threat to the native-born, middle class Americans, it also became symbolic of the destruction of the family and particularly to the status of women (Powers, 1991). Saloons were viewed as institutions that were almost anti-family, enticing patrons to mock traditional values (Lender & Martin, 1987, pp. 106-107).

These forces led to the ratification of the Eighteenth Amendment to the Constitution (the Volstead Act of 1919) which prohibited the manufacture, sale, transportation, and importing or exporting of "intoxicating liquors" from or to anywhere within the United States or territories under its jurisdiction. Much has been written about the failure of Prohibition, the "Great Social Experiment" to enhance positive social change (Miron, 1998; Schwartz, 1992; Tyrrell, 1997). It has even been argued that beyond the ideology that accompanied Prohibition, the rise of income tax as a revenue mechanism that far exceeded alcohol and customs taxes at that time provided more of an impetus for the support of Prohibition, and correspondingly, the subsequent drop in income tax during the depression led to Prohibition's repeal (Boudreaux & Pritchard, 1994). In fact, alcohol use, while reduced, did not end during Prohibition; yet, criminal activity around alcohol manufacture and sale increased dramatically. Moreover, Prohibition was a law that was essentially "unenforceable." Thus while the original intent of the law was to restore the social fabric, it further unraveled it in different ways. Beyond the direct effects of Prohibition, the Great Depression caused by the stock market crash in October 1929 contributed dramatically to the repeal of the Eighteenth Amendment. With an unemployment rate of almost one-third of the population by 1932, the government looked for ways to boost the economy. The Twenty-First Amendment that repealed the Eighteenth Amendment (Volstead Act) provided one direction towards economic recovery by developing more jobs and by providing new taxes on alcoholic beverages to boost the depressed economy.

The Development of the Disease Model:
Jellinek and *The Grapevine Study*

The general assumptions behind the "modern" disease model as exemplified in Alcoholics Anonymous can be traced as far back as the 1840s to the Washingtonian movement. This movement began when a group of six men in Baltimore, who recognized that alcohol was a negative influence in their lives, adopted the tenets of a local temperance society and took an abstinence pledge. Beyond their quest for abstinence, this movement included active participation and a willingness to help each other remain sober. This aspect of the program was considered a key element in maintaining sobriety. Help often involved assistance with finding jobs or temporary financial help. The Washingtonian movement hit its peak in the middle 1840s and signified a clear shift from social reform affecting an entire population to a clear focus on individual alcoholics with the primary goal of helping each other (Lender & Martin, 1987).

Even though the Washingtonian movement lost most of its support by the end of the 1840s, this shift from societal level of concerns to an individual level was to repeat itself historically several times before the AA movement officially started in the form of such groups as the Sons of Temperance, the Good Templars, and the Temple of Honor and Temperance. However, these groups were more enmeshed with the larger Temperance Movement and were not exclusively aimed at individuals who drank excessively. By 1919, the U.S. established Prohibition as a national alcohol policy by passing the Volstead Act. Thus, the need for Temperance societies ceased to exist. Upon repeal of Prohibition in 1933, the U.S. lacked any cohesive policy concerning alcohol (which some would argue is still true), and thus, treatment approaches varied.

The history of AA has been well documented (Kurtz, 1979; Whyte, 1998) and at least initially resembled the early Washingtonian movement. However instead of six, two men joined together to establish this fellowship: Dr. Robert Smith and William Wilson. The key element of their union was an agreement that they were powerless over alcohol and that they and other people like them had a disease called alcoholism. The founders of AA were also influenced by the Oxford Group, a movement of the 1920s and 1930s, which emphasized the healing of problems through spiritual change. The principles of the program were threefold including: (1) "four absolutes:" absolute honesty, absolute purity, absolute unselfishness, and absolute love; (2) "five C's:" confidence, confession, conviction, conversion, and continuance; and (3) "five procedures:" give in to God, listen to God's direction, check guidance, restitution, and sharing through witness (Dick, 1992; Whyte, 1998). Although the Oxford Group was not explicitly a treatment program for alcoholics and they did not require total abstinence from alcohol, many alcoholics were drawn to this program prior to the development of AA.

By 1939, Dr. Bob and Bill W. (their preferred names in AA) began to organize groups of alcoholics utilizing a set of principles for recovery. The criteria for joining AA was a desire to stop drinking and a willingness to help others. Anonymity was also stressed so that the groups focused more on the principles of AA but were also independent and hence not influenced by outside groups. By 1946, the Twelve Traditions were developed and adopted at the Alcoholics Anonymous International Convention in 1950 and were formally published as a book in 1953 (*Twelve Steps and Twelve Traditions*,

Table 3.1
Twelve Steps of Alcoholics Anonymous

1.	We admitted we were powerless over alcohol—that our lives had become unmanageable.
2.	Came to believe that a Power greater than ourselves could restore us to sanity.
3.	Made a decision to turn our will and our lives over to the care of God *as we understood Him.*
4.	Made a searching and fearless moral inventory of ourselves.
5.	Admitted to God, to ourselves, and to another human being the exact nature of our wrongs.
6.	Were entirely ready to have God remove all these defects of character.
7.	Humbly asked Him to remove our shortcomings.
8.	Made a list of all persons we had harmed, and became willing to make amends to them all.
9.	Made direct amends to such people wherever possible, except when to do so would injure them or others.
10.	Continued to take personal inventory and when we were wrong promptly admitted it.
11.	Sought through prayer and meditation to improve our conscious contact with God *as we understood Him*, praying only for knowledge of His will for us and the power to carry that out.
12.	Having had a spiritual awakening as the result of these steps, we tried to carry this message to alcoholics, and to practice these principles in all our affairs.

Source: (*Twelve Steps and Twelve Traditions* 1953)

Table 3.2
Twelve Traditions of Alcoholics Anonymous

One—Our common welfare should come first; personal recovery depends upon AA. unity.

Two—for our group purpose there is but one ultimate authority—a loving God as He may express Himself in our group conscience. Our leaders are but trusted servants; they do not govern.

Three—The only requirement for AA membership is a desire to stop drinking.

Four—Each group should be autonomous except in matters affecting other groups or AA as a whole.

Five—Each group has but one primary purpose—to carry its message to the alcoholic who still suffers.

Six—An A.A. group ought never endorse, finance or lend the AA name to any related facility or outside enterprise, lest problems of money, property and prestige divert us from our primary purpose.

Seven—Every AA group out to be fully self-supporting, declining outside contributions.

Eight—Alcoholics Anonymous should remain forever nonprofessional, but our service centers may employ special workers.

Nine—AA, as such, ought never be organized; but we may create service boards or committees directly responsible to those they serve.

Ten—Alcoholics Anonymous has no opinion on outside issues; hence the AA name ought never be drawn into public controversy.

Eleven—Our public relations policy is based on attraction rather than promotion; we need always maintain personal anonymity at the level of press, radio and films.

Twelve—Anonymity is the spiritual foundation of all our Traditions, ever reminding us to place principles before personalities.

Source: (*Twelve Steps and Twelve Traditions* 1953)

1953) (see tables 3.1 and 3.2).

Like the Washingtonian movement, these Steps and Traditions codified a treatment plan that was individualistic and involved ongoing participation and self-help. Similarly, groups provided a new social structure for alcoholics giving them camaraderie and friendship during meetings by creating a confessional group atmosphere. AA's strong investment in this new structure was explicitly meant to provide a replacement of former social networks that were far less supportive of an abstinent lifestyle.

While AA had considerable success in terms of attracting people, it lacked a scientific base for its contention that alcoholism is a disease. If alcoholism is a disease, then how is it diagnosed and what are the symptoms and the progression of these symptoms that eventually define alcoholism? To handle this "problem," E.M. Jellinek, a professor in applied physiology at Yale University, was recruited by the editors of *The Grapevine* (the official publication of Alcoholics Anonymous) in 1945 to analyze data from their own study of AA members in order to examine and develop a disease model of alcoholism that included a set of symptoms that were both identifiable and diagnosable. In 1946, the results of Jellinek's analysis of these data were published in an extensive article entitled "Phases in the Drinking History of Alcoholics. Analysis of a Survey conducted by the Official organ of Alcoholics Anonymous" in the *Quarterly Journal of Studies on Alcohol.*

Essentially this *"Grapevine Study"* (as it has been called) involved an anonymous, mailed survey of approximately 1600 AA members who received the *Grapevine* in May, 1945. The survey consisted of thirty-six items (table 3.3) representing specific incidents or experiences that may have occurred during the respondents' drinking histories prior to joining AA. Respondents were instructed to complete the questionnaires in the following way:

> The purpose of this questionnaire is to ascertain at what age the incidents or experiences listed below first happened. The order in which they are set down may not accord with your own experience, but please fill in the year after the item anyway. If you never had a particular experience, leave the space blank. The samples are intended to be suggestive only and are in no way definitive. (Jellinek, 1946, p. 3)

While it is not clear from Jellinek's (1946) article how these items were specifically selected, we can presume that there was some type of a consensus among the leadership of AA at the time as to what constituted "appropriate" symptoms of alcoholism. However, it is

Table 3.3
The Grapevine Questionnaire

At What Age Did You First:

1. Get drunk? (No example or illustration is attempted or necessary. If you were ever drunk you will know what we mean) ....
2. Experience a blackout? (Example: Wake up in the morning after a party with no idea where you had been or what you had done after a certain point) ....
3. Start sneaking drinks? (Example: Take a quick one in the kitchen without anyone seeing you when you were pouring drinks for guests) ....
4. Begin to lose control of your drinking? (Example: Intend to have only a coupe and wind up cockeyed) ....
5. Rationalize or justify your abnormal drinking? (Example: Excuse your drinking on the ground that you were sad, or happy, or neither) ...
6. Attempt to control your drinking by changing its pattern? (Example: Deciding to drink only before dinner) ....
7. Attempt to control your drinking by going on the wagon? ....
8. Act in a financially extravagant manner while drinking? (Example: Cashing a check for more than you need and spending all of it without getting anything for it except a hangover) ...
9. Start going on week-end drunks? ....
10. Starting going on middle-of-the-week drunks? ....
11. Start going on day-time drunks? ....
12. Take a morning drink? (Example: Feel the need of and take a drink the first thing in the morning in order to get yourself going, or "for medicinal purposes only") ....
13. Starting going on benders? (Example: Staying drunk for more than a day without regard for your work or your family or anything else) ...
14. Develop indefinable fears? ....
15. Experience acute and persistent remorse? (Example. Realizing that you have made a fool of yourself while drinking without being able to shake the realization off) ....
16. Develop abnormal and unreasonable resentments? (Example: going into a rage because dinner wasn't ready the minute you got home) ....
17. Commit antisocial acts while drinking? (Example: Pick a fight with a stranger in a salon for no justifiable reason) ...
18. Realize that your friends and family were trying to prevent or discourage your drinking?
19. Become indifferent to the kind or quality of the liquor you drank so long as it did the business? ....
20. Experience uncontrollable tremors (i.e., the jitters, the shakes, or whatever your pet name is) after drinking? ....
21. Resort to taking sedatives to quiet yourself after drinking? ....

Table 3.3 (cont.)

23. Seek psychiatric advice or aid? (This includes advice or aid from any adviser, such as a minister, a priest or a lawyer, as well as from a psychiatrist)....
24. Have to be hospitalized as a result of drinking? ....
25. Lose a friend as the result of drinking? ....
26. Lose working time as a result of drinking? ....
27. Lose a job as the result of drinking? ....
28. Lose advancement in a job as the result of drinking? ....
29. Use alcohol to lessen self-consciousness concerning sex? ....
30. Attempt to find comfort in religion? ....
31. Desire to escape from your environment as a solution for the drinking problem? (Example: Deciding that all would be well if only you could get a job in Chicago instead of having to go on working in New York)....
32. Start solitary drinking? ....
33. Start to protect your supply? (Example: Buying a quart on the way home so you would be sure to have a drink in the morning) ....
34. Admit to yourself that your drinking was beyond control? ....
35. Admit to anyone else that your drinking was beyond control? ....
36. Reach what you regard as your lowest point? ....
    Please state the following: (a) Present age .... (b) Sex ....

Source: *(Jellinek 1946, 3-5)*

clear that there is somewhat of an *a priori* sequence to them. It begins with "lighter" symptoms such as getting drunk and sneaking drinks, and ends with more severe experiences such as hospitalization because of drinking and reaching one's lowest point. Thus, the anonymous authors of this questionnaire did have some strong preferences for what behaviors, experiences, or incidences were central to the progression and development of alcoholism.

To some extent, it is unfair to judge surveys done in the 1940s with survey research standards used today. Jellinek was in the same position as any researcher who chooses to conduct a secondary data analysis. The data had already been collected, and Jellinek apparently had not been consulted regarding the design of the study, the sample used, the mode of administration, or the content or design of the data collection instrument. He did, however, have his own "misgivings" and stated that...

> Statistical thinking should not begin after a survey or an experiment has been completed but should enter into the first plans for obtaining the data. In the questionnaire under consideration this requirement was neglected. (Jellinek, 1946, p. 5)

Yet, despite these limitations, Jellinek did choose to analyze the data in an attempt to provide some "systematization of the knowledge derivable from the drinking history" (p. 3) in order to develop what he termed "the phaseology of alcoholism." What is particularly interesting about his seminal article written in 1946 is that Jellinek to some extent justifies his work as very much needed because of the "practical disregard of the drinking behavior by psychologists and psychiatrists" (p. 3). Presumably, this research was his attempt to legitimize alcoholism as a disease separate from any psychological or psychiatric rubric. This approach is very much reflected in the wording of each item. There is an absence of concepts such as "defense mechanisms" or any other language that might signify alcoholism as a mental health issue.

Jellinek's (1946) rationale for who was included in the sample (members of AA) is quite revealing.

> There is, on the other hand, a great advantage in the fact that all subjects were members of Alcoholics Anonymous. It is generally known that it is difficult to get truthful data on inebriate habits, but there need be no doubt as to the truthfulness of the replies given by an A.A. to questions coming from his own group. This element of the survey is one of its greatest assets. (pp. 6-7)

While one has to admire Jellinek's trust of self-reports from AA members, it is revealing that he did not consider any biases that could have resulted from such a sample. Given that Jellinek was interested in determining the symptomatology of alcoholism as a disease, it might have been more appropriate to include in the sample both treated (and not just with AA) and untreated alcoholics/heavy drinkers. Within his article, Jellinek does point out the need for comparative data and struggles with the need to compare his findings from AA members with those of moderate drinkers of similar age, social, and economic backgrounds. In fact, Jellinek does propose to use data from a colleague, Dr. Anne Roe, who "informally interviewed a group of thirty-one scientists on a few aspects of their drinking habits" (p. 18). While admitting that this comparison group of Roe's is very small, it does provide him with some basis to compare what symptoms are exclusive to the alcoholic group. However, Jellinek does not seem to feel that the *Grapevine* data needed to be compared with additional samples of alcoholics and heavy drinkers who may or may not have received treatment services.

Despite the fact that the data were already collected, it might also have made sense for Jellinek to "step back" in the research process and to have conducted some in-depth interviews with a wide range of alcoholics and heavy drinkers (diverse in terms of ethnicity, gender, income, education, etc.) to determine if there were other aspects of alcoholism not included in the survey that were important to assess. Moreover, it may have also been useful to explore if there indeed is a definable symptomatology associated with alcoholism. In addition, Jellinek does not mention that the respondents may have been influenced by AA in terms of how they responded to the items. That is, there is always the possibility that the respondents might be fitting their own drinking histories in their responses to follow the prevailing AA philosophy concerning what events or factors preceded the start of their alcoholism. This way of responding to the items may have skewed the data in the direction of lending support for the AA model (Arminen, 1996).

In addition to these methodological problems not addressed by Jellinek (1946), a closer inspection of the 36 items indicates that there appear to be additional assumptions made about this disease prior to the analysis of the data. For example, there are several items that identify specific conditions that may or may not be related to excessive alcohol use. For example, items #14, #16, #22, #23 and #30 (table 3.3) all address thoughts or behaviors that, in the AA tradition, are attributed to excessive alcohol use. Yet, it is perfectly reasonable to assume that indefinable fears (#14) and abnormal and unreasonable resentments (#16) may have occurred for reasons that may or may not include alcohol use. For example, these experiences could have occurred as a result of depression. Yet, it appears that all negative experiences that occurred during a respondent's drinking history and prior to joining AA are attributed to the disease of alcoholism; this is clearly open to question.

This tendency to attribute all negative behaviors to alcohol use mirrors the reasoning used in the stories told within AA and included in their book often referred to as the "Big Book" or the "Bible"(A.A. World Services, 1939). Alcoholics, within this context, tell their stories primarily with reference to their use of alcohol. Positive experiences in their life are primarily expressed as anomalies or are couched in ways that relate how their excessive alcohol use turned positive into negative experiences. Some have referred to this phenomenon as the "reconstruction of history" implying that alcoholics in AA learn a specific framework in which to conceptualize and share their histories (Arminen, 1996). These histories are clearly depicted in terms of the before and after of their excessive alcohol use as well as the before and after of their arrival in AA.

> At twenty-five I had developed an alcoholic problem…My drinking habits increased n spite of my struggle for control. I tried the beer diet, the wine diet, timing, measuring, and spacing of drinks…I had a tough pull back to normal good health. It has been so many years since I had not relied on some artificial crutch, either alcohol or sedatives, Letting go of everything at once was both painful and terrifying. I could never have accomplished this alone. It took the help, understanding and wonderful companionship that was given so freely to me by my "ex-alkie" friends. This and the program of recovery embodied in the Twelve Steps. In learning to practice these steps in my daily living I began to acquire faith and a philosophy to live by. Excepted from "The Keys of the Kingdom" (A.A. World Services, 1955, p 305, 310-311).

> My drinking did not start until after I was thirty-five, and a fairly successful career had been established. But success brought increased social activities and I realized that many of my friends enjoyed a social drink with no apparent harm to themselves or others. I disliked being different so, ultimately, I began to joint them occasionally…Gradually the quantity increased, the occasions for a drink came more frequently; a hard day, worries and pressure, bad news, good news–there were more and more reasons for a drink…From the start I liked everything about the A.A. program. I liked the description of the alcoholic as a person who had found that alcohol is interfering with his social or business life. The allergy I could understand because I am allergic to certain pollens. Some of my family are allergic to certain foods. What could be more reasonable than that some people, including myself, were allergic to alcohol? Excerpted from "It Might Have Been Worse" (A.A. World Services, 1955, p 383-384, 389)

The authors of the questionnaire also presumed that if the respondent sought medical advice or aid (#22), psychiatric advice or aid (#23), or attempted to find comfort in religion (#30), these acts were directly related to excessive alcohol use. While this may happen to individuals with alcohol problems, it also commonly occurs to individuals (with and without alcohol problems) for many other reasons.

In addition to these items that attribute specific behaviors solely to excessive alcohol use, item #32 (start solitary drinking) also has some specific inherent assumptions. One presumes that the norm of the individuals (probably male) who developed this questionnaire was group drinking or drinking with ones family. Drinking alone conjures up a specific image of someone who has put alcohol in the center of his/her life and ultimately has no other social contacts. This scenario may in fact describe some types of abusive alcohol use, but it may or may not be a defining symptom for individuals who, for example, live alone and/or have limited social contacts.

As mentioned earlier, another interesting characteristic of these thirty-six items of the *Grapevine Questionnaire* is that it does not include any psychiatric or psychological terminology. Kurtz (1979) has explained that...

> The true communication of experience by 'the language of the heart' was essential to Alcoholics Anonymous. By 'keeping it simple' in an area all too often marked by terminological obfuscation, the program of Alcoholics Anonymous offered assistance in confronting denials to its adherents— and even, to modern culture, an invitation to examine its own not dissimilar problem. (p. 193)

Only one item (#5) comes close to describing defense mechanisms such as rationalization, denial, and projection yet specific psychiatric concepts are never used. The "keeping it simple" philosophy also served to reinforce that AA was a fellowship and run by peers who are themselves alcoholics—not mental health professionals. Despite this lack of psychological or psychiatric terms in the *Grapevine Questionnaire*, Jellinek often applies this terminology in his 1946 article. For example, in discussing a few men in the sample who began with solitary drinking at early stages of their drinking history, he implies they had psychological problems and indeed labels them.

> These early solitary drinkers had traits which indicated that they were psychological deviants before they started drinking. No doubt they were "lone wolves" from the very beginning, perhaps shy, perhaps suspicious of others. Some students of alcoholism may be inclined to call them schizoid personalities (p. 71)

A final comment on The *Grapevine Questionnaire* is that there is no place in it to indicate if a respondent's experiences somehow differed from the items themselves. Because the symptoms were developed *a priori,* respondents essentially reported their presence (their age when it first occurred) or absence during their drinking histories. An implicit assumption with this research is that there is content validity. That is, it is assumed that this questionnaire covers the range of what is considered the disease of alcoholism and that this set of symptoms is the same for all alcoholics regardless of age, gender, ethnicity, or class. It might have been useful to have an open-ended items or some place on the questionnaire where respondents could include other experiences they may have had that contributed to their alcoholism.

In May 1945, The *Grapevine* was distributed to its AA membership with this anonymous questionnaire on the front of its newsletter. Of the approximately 1,600 questionnaires that were mailed, only 158 members responded yielding a very low response rate of about 10 percent. Jellinek, not necessarily dismayed by such a small response rate, surmised that one reason for this low response rate may have been how the questionnaire was distributed.

> Such a small return gives rise to surmises on the possible selectiveness of the sample. On this score, however, I would be inclined towards optimism. The questionnaire was not printed separately but on the front page of the *Grapevine*. Since at that time group subscriptions predominated, individual members of any such group could not very well deprive other members of two pages of the magazine. This feature alone would tend to bring about a great reduction in the number of possible returns (p. 6)

In his final analysis Jellinek used only ninety-eight of the questionnaires. Of those that were dropped, fifteen were completed by female alcoholics. Jellinek's reasoning for not including women was twofold. First, with the small number it would be difficult to compare men and women. Second, he also apparently could see that the data from women were very different than men and thus felt it was not appropriate to merge the two together. Unfortunately, the data on women in The *Grapevine* Study were never analyzed. One wonders how the model might have been different if a careful analysis of the data provided by women alcoholics had been undertaken and somehow incorporated in the model.

Jellinek was keenly aware of many of these methodologic problems, and he states throughout his article that his findings are suggestive and not definitive. Presumably due to some frustration with his analysis, he even developed a revised version of the questionnaire that he discusses and includes it as an appendix to his 1946 paper. This revised questionnaire was analyzed and presented in a second article he published on the symptoms of alcoholism in 1952 (discussed in more depth below). In this second paper, Jellinek suggests separate versions for men and women "to avoid awkwardness in the formulation of questions" (p. 79) but, it is unclear if there will be different content items by gender.

Jellinek's (1946) analysis of the data collected in the *Grapevine* study was laudable given the limited capabilities researchers had at that time to analyze data. Throughout this eighty-eight-page article, Jellinek struggles to conduct a careful and painstaking analysis in an attempt to put these thirty-six symptoms in some kind of a reasonable and useable sequence in his construction of a coherent symptomatology so that they resemble a disease. His major frustration is with the lack of items that focus on "prealcoholic" drinking.

> Not every culturally deviant drinking behavior or reaction to alcoholic beverages denotes alcoholism...As they stand, the drinking histories of the *Grapevine* shed little

light on a possible preparatory phase of alcoholism, but they suggest a crucial or basic phase which may be valid for alcoholism in general. (p. 21)

Despite this limitation, Jellinek did manage to categorize most of these symptoms into what appears to be five phases of drinking of alcoholics: Basic Phase of Alcoholism, A tentative Intermediate Phase (second point of orientation), Prodromal Symptoms, Compulsive phase, and Terminal Phase. Interestingly, Jellinek wrote about these phases in a very tentative and somewhat unclear way presumably because he viewed these data as preliminary and limited. Throughout his analysis, he struggles to make sense of the ordering of the symptoms given the small sample size. In the Basic Phase of Alcoholism, Jellinek describes a progression of frequent use of alcohol, blackouts, and loss of control. A compulsion to drink defined by items related to midweek drunks, daytime drunks and "benders" characterized Jellinek's Intermediate Phase. In the third phase, Prodromal Symptoms, Jellinek included "blackouts" and "sneaking drinks." The Acute Compulsive Phase is defined primarily by two symptoms: going on the "water wagon," and "changing ones drinking pattern." Both are attempts to stop or otherwise avoid what has previously been defined as a "loss of control" over alcohol. Finally, the Terminal Phase is seen as the final stage of the downward spiral of the disease and is defined as "reaching the lowest point" – including "admitting to self that drinking was beyond control" and "admitting to anyone else that drinking was beyond control."

In Jellinek's (1952) second paper, he presents his phases of alcohol addiction once more but this time in a more systematic way. Within this article (originally published as an appendix to a World Health Organization report), he states that these results are based on a larger sample of 2,000 male alcoholics who responded to a revised questionnaire (forty-three items) that now included more items on "prealcoholic" drinking as well as more symptoms in general.

As previously noted, it may be unfair to compare very early survey research with contemporary standards. However, it is important to keep in mind that Jellinek had the opportunity to revise the original *Grapevine* questionnaire and include more items (the second questionnaire had 111 items, sixty-eight of which were directly related to the respondent's drinking behaviors) as well as revise those items that he felt were unclear in the original instrument. However, an analysis of the new items on the second questionnaire points to additional methodological problems.

Some of the newer items that Jellinek included are what survey researchers call double- or even triple-barreled questions. This means that they include several components that may or may not occur simultaneously. For example,

48. Age when you began drinking more than once a week but not every day, getting drunk only sometimes, and without any difficulties the next day.

96. Age when you began to have feelings of fear without knowing what you were fearing, or fearing that there might be retribution because of your excessive drinking.

In both of the above examples, the conditions necessary for giving a positive response are confusing and overly complex. For example, how would someone answer #48 if he or she met the drinking criteria, but never got drunk, and did not have any difficulties? Would the answer be "never" or would he or she write in an age?

In addition to the double- or triple-barreled items, there are also several additional questions that identify specific conditions or experiences that may or may not be related to excessive alcohol use. For example,

83. Age when you began to have ideas of jealousy concerning your wife or girl friend.

85. Age when you began suffering from sleeplessness.

86. Age when you began to pity yourself (feeling that everybody was down on you, that you deserved a better fate, that the world didn't give you a chance, etc., etc.)

87. Age when you thought the best solution would be to be dead.

88. Age when you first contemplated suicide.

90. Age when you sought or accepted the services of an intermediary to straighten out matters with your family, friends or employer.

97. Age when you began to having periods of despondency.

107. Age at which you began to feel a religious need.

Like some of the items in the original *Grapevine Questionnaire*, these experiences could have occurred as a result of excessive alcohol use or for other reasons such as depression or a need for structure. The continual attribution of all negative feelings and behaviors to alcohol use is a strong characteristic of AA principles and consequently of both the original and revised questionnaires.

Like the original questionnaire, there is an absence of any psychological or psychiatric terms are used to describe behaviors. However, the following two items do distinctly describe the use of rationalization and projection.

74. Age when you began to justify to yourself, or to find alibis for, your excessive drinking. (*Example*: Convincing yourself that you were fully able to control your drinking and that whenever you got drunk it was because of some good reason and not because of lack of control, or that your efficiency required alcohol, or that alcohol was a medicine for you "nervousness," and that there was no better medicine for you, etc., etc.)

79. Age when you began convincing yourself that any neglect to which you may have exposed your family was justified because your drinking was necessary for you or because "it was coming to them."

Finally, like the original questionnaire, Jellinek's revised questionnaire does not allow any input from the respondents regarding how their experiences may have differed from the ones presented in the questions. A marked improvement in the format of the questionnaire is that the there are three answer categories for each item that allows respondents to report a symptom even if they cannot remember the age at which it first occurred.

In contrast to his 1946 article, in this later study, Jellinek (1952) does not report on how the data were obtained from AA members. We are not told anything about the sample selection, original sample size, or even where the questionnaire was distributed. We do, however, know from the article that once again the data are only from male members of AA. This time Jellinek's reasoning is less apologetic and more methodological. "For alcoholic women, the "phases" are not as clear-cut as in men and the development is frequently more rapid" (p. 676). This apparently provided the rationale for eliminating women from the phases. A different approach might have been to take a "female normative" position arguing that the male alcoholics showed a curiously slower development of each of the phases and were more rigid. Nonetheless, while the sample size of the second study is larger, it still represents men only, and it is not clear if there is any age or ethnic diversity in this new sample.

In sharp contrast to his earlier article, the lack of any information of how Jellinek analyzed his data for the second study is striking. There are no data on the average age at which the sample reported any of the symptoms, for example. The data were not included in any article of Jellinek's nor in his 1960 book entitled *The Disease Concept of Alcoholism* (Jellinek, 1960). In addition, there is a lack of information about the study design and sample size. This has led some researchers to question whether the larger data set ever existed and/or whether it had ever been analyzed (Room, 1978). Yet, despite these substantial omissions, Jellinek presents four very distinct phases: the Prealcoholic Symptomatic Phase, the Prodromal Phase, the Crucial Phase, the Chronic Phase. Within each of these four phases are a series of symptoms. "Loss of Control", the eighth symptom, is clearly the critical point that signifies the transition from the Prodromal Phase (symptomatic) to the Crucial Phase (addictive).

These phases became the central core of the alcoholism treatment, or as Room (1978) has termed it—the *locus classicus* of the disease concept of alcoholism. This is exemplified by the famous "Chart of Alcohol Addiction and Recovery" printed in the 1958 volume of the *British Journal of Addiction* by M.M. Glatt (figure 3.1). This chart represents a melding of Jellinek's "Chart of Addiction" with the experiences of recovered ex-patients from his alcoholism treatment program at Warlingham Park Hospital in Britain. While not all of Jellinek's symptoms are mentioned in Glatt's chart, those that are included are placed in the sequence described in Jellinek's 1952 article.

Figure 3.1
A Chart of Alcohol Addiction and Recovery

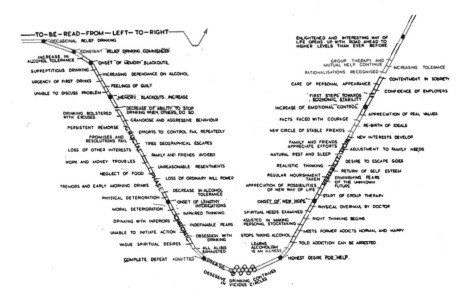

Source: Glatt, 1958

What is probably one of the most interesting findings from assessing both Glatt's and Jellinek's "charts" is that at least some of the symptoms, particularly those that focus on moral "behavior," do not appear to come from the items included in either questionnaire. For example, in both charts, reference is made to "marked ethical deterioration—symptom #32" (Jellinek, 1952) or "moral deterioration" (Glatt, 1958) yet it is not clear which items, if any, were used from the revised questionnaire to define this concept. This is also true for "drinks with persons far below his social level - symptom #35" (Jellinek, 1952) and "drinks with inferiors" (Glatt, 1958). This issue of social class may say more about the Jellinek's perception of social roles than what actually occurs during the latter stages of alcoholism. Consequently, this "liberty taken" by Jellinek calls into question how much leeway Jellinek allowed himself to fit specific symptoms into his conceptualization of alcoholism whether or not they were included in his revised questionnaire.

Assumptions Behind the Disease Model

Not discussed in this chapter are other models of alcoholism that co-exist with the disease model. Beyond the "old" and "new" medical models and the AA model discussed by Siegler, Osmond, and Newell (1968), they also describe five additional models that include the "impaired" model, the "dry" moral model, the "wet" moral model, the psychoanalytic model, and the family interaction model. While there are additional models, when alcoholism is viewed from an individualistic perspective, all of these models can be seen to operate under the same set of assumptions:

(a)   that the phenomena subsumed have enough in common to be regarded as an entity;

(b)   that the entity involved in a condition rather than an occurrence— that is, its manifestations are continuous or repetitive;

(c)   that the condition is abnormal (if not in actual fact and statistically, at least in comparison to some idealized criterion); and

(d)   that it is undesirable—something should be done about it. (Room, 1978)

Room (1978) has further suggested that there are three basic assumptions that form the basis of the disease model of alcoholism as originally developed by Jellinek. The first assumption is *unilinearity* defined as one set of stages through which all alcoholics must complete. Jellinek's focus on the sequencing of specific behaviors or experiences makes this assumption explicit. This "one-size fits all approach" of the disease model has been debated in the literature (Fingarette, 1988; Room, 1983) yet the main stages as outlined by Jellinek continue to be used as the main trajectories for becoming alcoholic. The second assumption is *accretionality,* which refers to a type of building block structure whereby each stage builds upon the stage before it. Further, once an alcoholic reaches a particular stage, there is a perma-nence to it with no opportunity at any time to reverse the stage. This either/or approach allows for the use of a "recovery" approach to treat-ment as opposed to a "cure" orientation since it is assumed that once one becomes an alcoholic, this condition is intractable and remains a lifelong disease. The last assumption is *immanence.* This assumption refers to the inner state and behavior of the person him/herself. It im-plies that the problem, alcoholism, came from inside the person rather than being a reaction to environmental or outside circumstances.

Room (1978) tested the assumptions of universality of symptoms and of sequencing of symptoms by using data from Jellinek's origi-nal study along with studies of people labeled as "alcoholics" from other sources. If the disease model is indeed a universal phenom-enon, then individuals who are considered alcoholics should expe-rience similar symptoms in a similar sequence. Room (1978) found that this was not the case. As expected, the AA samples had the highest prevalence for any specific item as compared to hospital/ clinic samples. This casts doubt on these symptoms universality. Moreover, the time-ordering of these symptoms based on age of onset do not fit into a clear sequence - including the crucial symp-tom "loss of control." In his later work, Jellinek (1960) even con-ceded that probably 10-15 percent of AA members had a pathology that preceded alcoholism (which he termed alpha alcoholism) and that these members probably changed the way they described their alcoholism so that it fit into the AA philosophy. Regardless of the inability to demonstrate empirically its universal applicability, the disease model experienced a tremendous amount of growth and expansion since its inception in the late 1930s.

The Growth of the Disease Model:
Al-Anon, ACAs, and Co-Dependency

Al-Anon

Perhaps one of the most intriguing aspects of the AA movement and the 12-step tradition is the diffusion of this structure to other populations seeking help. Early on in the AA movement, alcoholics and their families went to group meetings together, but this was stopped sometime during the 1940s when the meetings grew too large (Whyte, 1998), and there was strong support for AA meetings to be exclusive of family members. Thus, family member groups began to meet separately and were primarily composed of wives of alcoholics whose spouses were members of A.A. The groups themselves took on several forms and names such as Non-A.A. Group (NAA), Triple A, AA Associates, AA Auxiliary, AA Helpmates, Alono, and Onala (Whyte, 1998). In March 1952, Al-Anon, an abbreviation of Alcoholic Anonymous that combines the first syllables of each word, became the name of this organization of AA family members seeking their own group.

In *Lois Remembers* (Lois W., 1979), Lois W., the co-founder of Al-Anon and the wife of Bill W., recalls in detail the steps Al-Anon took to create an organization that closely paralleled Alcoholics Anonymous. Their first task was to decide on the basic principles by which each group operated. Some groups had already developed their own principles but over time, all groups "...recognized the power of AA's Twelve Steps, even in the wording itself, and we adopted them for our own guidelines" (Lois W., 1979; p. 176). In addition, Al-Anon also adopted AA's Twelve Traditions. No doubt, Lois W. was influenced by her husband and the AA philosophy. It is interesting to wonder how Al-Anon may have been developed had it not been for this strong influence.

Like AA, Al-Anon published its own directory of meetings and also published its own newsletter and other materials explaining its nature and purpose. In 1953, Lois W. spoke on television about Al-Anon and several articles were written about it in the popular press (e.g., *Christian Herald*, *Life Romances*, and *Life Today*). Also, similar to AA, an article appeared in the *Saturday Evening Post* in July 1955 that further served to provide exposure of the Al-Anon organization to a mass audience. Over time, Al-Anon's "story" was told many times in articles, books, and on television which "...enhanced public awareness of Al-Anon and brought many new members into the groups" (Whyte, 1998, p. 221).

To a large extent, the assumptions of Al-Anon concerning the disease model of alcoholism very much parallel those of AA. As stated earlier, the Twelve-Steps and the Twelve Traditions are almost identical with only one difference. Al-Anon's twelfth step reads "carry this message to *others*" instead of "to alcoholics."

Yet, in addition to the inherent tenets Al-Anon shares with AA, there are an additional set of assumptions about the boundaries of this disease that are important to note. Al-Anon adheres to the philosophy that alcoholism is a family disease affecting partners, children, relatives, etc. in such as way as to also render them "diseased" in some fashion. This to a large degree expands the eligible potential treatment population to include those "auxiliary" people in the alcoholic's life.

In recent years, there has been a major growth in groups modeled on AA and de facto, Al-Anon. A partial list of these anonymous self-help organizations in addition to AA includes: Batterers Anonymous, Co-Dependents Anonymous, Debtors Anonymous, Depressives Anonymous, Divorce Anonymous, Drugs Anonymous, Emotional Health Anonymous, Emotions Anonymous, Emphysema Anonymous, Families Anonymous, Fundamentalists Anonymous, Gamblers Anonymous, Grandparents Anonymous, Homosexuals Anonymous, Impotents Anonymous, Incest Survivors Anonymous, Marriage Anonymous, Molesters Anonymous, Messies Anonymous, Narcotics Anonymous, Neurotics Anonymous, Overeaters Anonymous, Parents Anonymous, Potsmokers Anonymous, Prison Families Anonymous, Sex Addicts Anonymous, Sex and Love Addicts Anonymous, Workaholics Anonymous (Rice, 1996, p. 51).

In addition, Rice (1996) lists the following variations of these groups: Cocaine Anonymous, Men for Sobriety, Pills Anonymous, Shoplifters Anonymous, Smokers Anonymous, Spenders Anonymous, Survivors of Incest Anonymous (primarily male victims), Twelve Steps for Christian Living Group, Twelve Steps for Spiritual Growth, Women for Sobriety, Women with Multiple Addictions (pp. 51-52). The general assumption behind these twelve step groups is that there is both a primary and secondary disease. The primary disease belongs to the addict him/herself, for example, the cocaine abuser, the alcoholic, the smoker. The secondary disease or secondary problems are assigned to the partners, family, children, friends and other associates of the person who has the disease. Rice (1996) argues that for almost all the "anonymous" groups there are parallel secondary groups in the subculture that often refer to themselves as "co" plus whatever the disease is. To understand better this phenomenon of co-dependency, it is first necessary to discuss the development of the Adult Children of Alcoholics (ACA) Movement and its impact on the recognition of a treatment subculture in the U.S.

Adult Children of Alcoholics (ACAs) and the Emergence of Co-Dependency

The focus on children (youth) and alcoholism is not a new one. The negative effects of alcohol on the family and particularly on children was portrayed during the temperance era in the nineteenth and early twentieth century by songs, plays, best sellers, editorial cartoons, newspapers and journals (O'Gorman, 1985). With the repeal of prohibition, the interest in the children and of the family dynamics specific to alcoholism lost its popularity. Consequently, treatment issues and research were aimed at the alcoholic him or herself.

The re-emergence of an interest in children of alcoholics was fostered by mental health professionals working with a new adult client population who self-identified as Children of Alcoholics (COAs) or Adult Children of Alcoholics (ACAs) (Hunt, 1989) as well as several other forces. First, the focus of AA, Al-Anon and Alateen (a group based on Al-Anon principles designed for adolescents) on the disease not only of the individual but also of his or her relationships invoked a need for a group who were previously directly affected by the "disease" but are currently being affected by it. Second, state and federal agencies have actively provided funds for services related to the adult children of alcoholics. Moreover, national conferences have been held and publications have emerged that focus directly on children of alcoholics and their issues. Finally, the National Association for Children of Alcoholics (NACoA) was developed in 1983 whose mission is to provide information and services for children of alcoholics and for those treatment professionals who work with this population (Rudy, 1991).

O'Gorman (1985) contends that 1979 was the beginning of informal discussions by such professionals which would lead to a nationwide movement to alleviate the perceived sufferings of COAs. Sharon Wegscheider, Claudia Black, Janet Woititz were invited by Jean Kroc to the Kroc Ranch in California [Krocs of McDonald's fame] to discuss this new treatment group. From this meeting and several others, there was a general recognition of the problem similarities reported by ACA's. Utilizing a family systems perspective with the alcoholic family, Wegscheider noticed consistent roles and relationships that would later be labeled "co-dependency." Eventually, groups started to be formed for this "newly discovered group" and virtually without any formal research, books and pamphlets were being published focusing on the 'dysfunctional nature of alcoholic family systems and the special problems of COAs. Along with NACoA being founded in 1983, the Children of Alcoholics Foundation was created in 1984 that called for research and education on alcoholism and its effects of children. Eventually, regional conferences were convened, and the first national conference was held in Florida in 1984. Examples of some of the topics of the regional conferences, aimed at both professional and lay people are: alcoholism as a family disease; the high risk of COAs for developing alcoholism [cited as three to five times higher]; the prevalence of physical and sexual abuse/incest within alcoholic homes; treatment issues such as problems of intimacy associated with ACAs; and, rigid role performances learned within alcoholic homes (Hunt, 1989).

The stated goals of the movement were: (1) to encourage COAs to form self-help groups; (2) to inform practice of the therapeutic needs of this special population; and (3) to encourage researchers to utilize this population in future research endeavors. This new movement was different from the traditional Al-Anon philosophy in that it emphasized the disease of co-dependency and conceptualized this disease as similar to the Glatt V-chart (figure 3.1) in terms of ones decline and ones recovery process from this disease.

The ACA movement was very attractive to those individuals attempting to understand problems in their current life in terms of a family of origin which included alcoholism (Room, 1987). The ACA literature includes specific role-types that ACAs may have acquired as a result of their alcoholic homes. For example, an ACA may have taken on the *child-hero* role defined as an overachiever motivated by inadequacy and guilt. Additional roles include the *scapegoat* (delinquency, acting out motivated by hurt), the *lost child* (shyness, solitariness, motivated by loneliness), and, the *mascot* (clowning, hyperactivity, motivated by fear) (Black, 1981; Wegscheider, 1981).

As is true for groups utilizing the Twelve Step framework, a reexamination or an inventory of ones life is a necessary part of the recovery process. For ACAs, this exercise involves not only looking at ones own drinking patterns but also more broadly focusing on any addictive relationship patterns that stems from this tendency towards co-dependency. Room (1987) cites movement-related books such as Janet Woititz' *Struggle for Intimacy* (Woititz, 1985) and Robin Norwood's *Women who Love too Much* (Norwood, 1985) as examples of its broad based approach or what Rice (1996) terms "liberation psychology" that provides a ready-made cause to any and all current relationship "ills."

The umbrella term of "co-dependency" began to be used more extensively in the 1980s to explain dysfunctional adult behavior stemming from family of origin alcoholism problems. Co-dependency (originally termed "co-alcoholic") was first coined in the Ala-Anon movement but became more widely used and adopted by those in ACA. Because it is so widely used in the literature and in treatment groups and has become a large part of the popular self-help culture, it is important to understand how it is defined and used.

The literature is divided by those who endorse the co-dependency concept as an important element of Twelve Step life (e.g., Wegscheider-Cruse, 1985) and those who are skeptical of this amorphous term (e.g., Rice, 1996). It has even been argued that co-dependency should be a diagnosable condition that warrants a formal diagnosis similar to Alcohol Abuse or Alcohol Dependence (Cermak, 1984). Critics, however, argue that its very definition is unclear and often tautological. If its definition is applied to anyone who is affected by a substance user (Larsen, 1983), this would include a wide range of individuals including spouses, lovers, children, friends, etc. who could claim the label (or "status" as the case may be) of co-dependent. Thus, the label of co-dependent could be applied to virtually everyone (Gomberg, 1989).

Asher (1992) defines co-dependence, like alcoholism, as a social construct that needs to be understood in relation to the larger framework of the labeling and medicalization of deviance. The co-dependence label, like the disease concept of alcoholism, is medicalized thereby implying the need for treatment and recovery for those affected by the alcoholic. Yet, one cannot talk about the co-dependency movement without acknowledging extremely important gender issues that strongly influence its widespread appeal. Most of those individuals who consider themselves co-dependent are women living with alcoholic husbands. Thus, there is always the potential of using the label "co-dependent" to punish women further for behaviors they were socialized to perform (Asher, 1992).

> I think we need to be keenly aware of the relationship between conceptualizations of codependency and historical attitudes and actions toward women. There has been and still is a staggering flow of cultural messages about women's second-class citizenship, such as women's vulnerability, subservience, caretaking, passivity, self-sacrifice, sexual

degradation, and affective and economic dependence. There are burdensome social expectations for women to keep families together, and to bridge and smooth discrepancies between private and public family life. Until recently, there has been a virtual absence of cultural mandates or socially structured opportunities for females to develop resources and skills of emotional and economic autonomy, assertiveness, and psychological or social empowerment. All of this fits disconcertingly well with the characterization and implications of codependency. Given this cultural context, I believe it is primarily women who are label codependent, and it is primarily female spouses of alcoholics who experience life with their mate as a moral career. Much of what can be said about female socialization and experience also pertains to codependency, and those fundamental experiential aspects of a woman's life predate any later encounter with a moral career of being married to an alcoholic. (Asher, 1992; 197)

Given the pervasiveness of the ACA/Co-dependency movement which began in the 1980s and continues to have a pronounced influence in the U.S. today, it is important to examine its timing. That is, why did this movement evolve during this time period and not earlier particularly because the twelve-step movement (AA) developed almost sixty years ago? Room (1987) has reasoned that the emergence of this movement is generational. That is, the founders of this movement, members of the "baby boom" generation, came of age in the 1960s during which time alcohol consumption in the U.S. was rising rapidly but basically ignored due to the concern over illicit drugs. Some of these baby boomers may have aged out of Alateens, a Twelve Step group for children of alcoholics. In addition, during that time period, Room (1987) contends that new gender roles were being created that may have led to different types of relationship issues. A second explanation is that the movement may be seen as a response to the costs to women of men's drinking. In general, ACA members are more likely to be women who had a male alcoholic relative (usually a father). The term and perhaps label of "co-alcoholic" may be seen as somewhat empowering to the ACA members by providing legitimacy to their position as "innocent" and not responsible for their current life problems. It is not surprising that the movement was founded by women therapists primarily working with women clients. Finally, ACA's popularity may also be connected to its focus on self-control—on taking control of ones life—which coincides well with American culture and values. This co-existed very well with the political philosophy of 1980s. The Reagan era, sometimes referred to as the "New Right," included severe cutbacks in spending for social services programs. Thus, a movement that emphasized self-help groups also provided a way to conveniently fill the widening gap for the services that had previously been funded.

Summary

This chapter has outlined the rise of the disease model of alcoholism and its expansion to other groups of individuals associated with the alcoholic. By putting excessive drinking and other deviant behaviors under the growing purview of medicine, a whole new array of definitions and treatment options became available under medical control without many legal and procedural restraints. Yet, its association with the medical profession also served to provide alcoholism with some level of status, now making it a "legitimate" illness that warranted a medical diagnosis and treatment. Jellinek's *Grapevine Study* provided the research credibility to make such claims and to integrate this "new" disease more fully into mainstream American life and culture.

Jellinek's work endorsed the position that alcoholism as a disease was based on a series of progressive steps that were irreversible and once diagnosed, required a lifetime commitment to recovery through abstinence and attendance at Twelve Step groups. The growth of this movement into other parallel groups was enormous, and these other groups quickly became integrated into the U.S. culture. In particular, codependency as a condition and as a lifestyle achieved a widespread appeal and acceptance as an important treatment area.

With the history of the disease model and the medicalization of deviance comes the basic assumptions about human behavior that underlie these movements. Repeatedly, AA stories reflect individuals who go to desperate lengths to avoid admitting that their drinking is out of control and that their alcohol use is the cause of many if not all negative outcomes such as job loss, loss of family, health problems, etc. It is assumed that coming to terms with the disease and starting ones recovery can only begin when individuals recognize that they are alcoholics and that their alcoholism is the cause of many unwanted events.

While this discussion has focused on shifts in ideology as manifested by social movements within the alcohol field, institutional shifts have also reinforced not only the process of medicalization in the alcohol field but also more recent shifts towards the biomedicalization of alcohol studies. This in turn further drives the discourse towards search for the etiology, diagnosis and treatment of the "problem" within the individual. Chapter 4 will describe the process by which the National Institute on Alcohol Abuse and Alcoholism (NIAAA), along with the National Institute on Drug Abuse

(NIDA) and the National Institute of Mental Health (NIMH) moved their research mandates from the Alcohol Drug Abuse and Mental Health Administration (ADAMHA) to the National Institutes of Health (NIH) in 1992. By examining this institutional shift by addressing the materials produced at that time, I explore the rationale for this move as well as some of the concerns voiced about the possible problems posed by such an organizational and ideological shift. This move to NIH codified, legitimated, and intensified NIAAA's primary focus on the "disease" model of alcoholism and moved it further away from a broader range of what had historically also been considered salient issues in the alcohol field.

References

A.A. World Services. (1939). *Alcoholics Anonymous: The Story of How Many Thousands of Men and Women Have Recovered from Alcoholism*. New York City.

A.A. World Services. (1955). *Alcoholics Anonymous: The Story of How Many Thousands of Men and Women Have Recovered from Alcoholism* (Second edition ed.). New York City.

Arminen, I. (1996). On the Moral and Interactional Relevancy of Self-Repairs for Life Stories of Members of Alcoholics Anonymous. *Text, 16*(4), 449-480.

Asher, R. M. (1992). *Women with alcoholic husbands: ambivalence and the trap of codependency*. Chapel Hill: The University of North Carolina Press.

Black, C. (1981). *It Will Never Happen to Me!* Denver, CO: MAC.

Boudreaux, D. J., & Pritchard, A. C. (1994). Price of prohibition. *Arizona Law Review, 36*(1), 1-10.

Cermak, T. L. (1984). Children of Alcoholics and the Case for a New Diagnostic Category of Codependency. *Alcohol Health and Research World, 8*, 38-42.

Conrad, P., & Schneider, J. W. (1992). *Deviance and medicalization: From badness to sickness*. Philadelphia: Temple University Press.

Dick, B. (1992). *The Oxford Group & Alcoholics Anonymous*. Seattle, WA Glen Abbey Books, Inc.

Fingarette, H. (1988). *Heavy drinking: the myth of alcoholism as a disease*. Berkeley: University of California Press.

Glatt, M. M. (1958). Group therapy in alcoholism. *The British Journal of Addiction, 54*(2), 133-148.

Gomberg, E. L., S. (1989). On Terms Used and Abused: The Concept of 'Codependency'. *Drugs and Society, 3*(3/4), 113-132.

Goode, E. (1999). Theories of Drug Use. In E. Goode (Ed.), *Drugs in American Society* (Fifth ed., pp. 91-118). San Francisco, CA: McGraw-Hill.

Grob, G. N. (1981). *Nineteenth-Century Medical Attitudes Toward Alcoholic Addiction*. New York: Arno Press.

Hunt, M. (1989). *Children of alcoholics: family of origin factors and adult outcomes among the middle generation of a three-generation longitudinal panel*. Unpublished Ph.D. dissertation, Social Welfare, University of California, Berkeley.

Jellinek, E. M. (1946). Phases in the drinking history of alcoholics: Analysis of a survey conducted by the Official Organ of Alcoholics Anonymous. *Quarterly Journal of Studies on Alcohol, 7*, 1-88.

Jellinek, E. M. (1952). Phases of alcohol addiction. *Quarterly Journal of Studies on Alcohol, 13*, 673-684.

Jellinek, E. M. (1960). *The disease concept of alcoholism*. New Bruswick, NJ: Hilhouse Press.

Keller, M. (1990). *Models of Alcoholism. From Days of Old to Nowadays*. New Brunswick, NJ: Center of Alcohol Studies, Rutgers University.

Kurtz, E. (1979). *Not God: A history of Alcoholics Anonymous*. Center City, MN: Hazelden.

Larsen, E. (1983). *Basics of Co-Dependency*. Brooklyn Park, MN: Larsen Enterprises.

Lender, M. E., & Martin, J. K. (1987). *Drinking in America: a history* (The revised and expanded ed.). New York: The Free Press.

Levine, H. G. (1978). The discovery of addiction: changing conceptions of habitual drunk- · enness in American history. *Journal of Studies on Alcohol, 39*(1), 143-174.

Lois W. (1979). *Lois Remembers: Memoirs of the co-founder of Al-Anon and wife of the co-founder of Alcoholics Anonymous*. New York: Al-Anon Family Group Headquarters, Inc.

Miron, J. A. (1998). Economic analysis of alcohol prohibition. *Journal of Drug Issues, 28*(3), 741-762.

Norwood, R. (1985). *Women Who Love Too Much*. Los Angeles, CA: St. Martin's Press.

O'Gorman, P. (1985). An historical look at children of alcoholics. *Focus on the Family, 8*(1), 5-6.

Powers, M. (1991). *Faces among the bar: lore and order in the workingman's saloon, 1870-1920*. Unpublished PhD dissertation, University of California, Berkeley, Berkeley.

Rice, J. S. (1996). *A Disease of One's Own*. New Bruswick, NJ: Transaction Publishers.

Room, R. (1978). *Governing Images of Alcohol and Drug Problems: The Structure, Sources and Sequels of Conceptualizations of Intractable Problems*. Unpublished PhD dissertation, University of California, Berkeley, Berkeley.

Room, R. (1983). Sociological aspects of the disease concept of alcoholism. In *Research Advances in Alcohol and Drug Problems* (Vol. 7, pp. 47-91). New York: Plenum Press.

Room, R. (1987). *Changes in the cultural position of alcohol in the United States: the contribution of alcohol-oriented movement*. Paper presented at the Per une sociologia dell-alcool: Confronto internaziionale sui modelli del bere nel mutanto sociale (Alcohol and social science: an international forum on drinking patterns in relation to social change., Torino and San Stefano Belbo, Italy.

Rudy, D. R. (1991). The adult children of alcoholics movement: A social constructionist perspective. In D. J. Pittman & H. R. White (Eds.), *Society, Culture, and Drinking Patterns Reexamined* (pp. 716-732). New Brunswick, New Jersey: Rutgers Center of Alcohol Studies.

Rush, B. (1814). *An inquiry into the effects of ardent spirits upon the human body and mind, with an account of the means of preventing, and of the remedies for curing them*. Brookfield, MA: Reprinted in Grob, G (1981). See above reference.

Schwartz, R. H. (1992). Prohibition, 1920 to 1933: An overview of its effects on public health and the economy. *Southern Medical Journal, 85*(4), 397-402.

Siegler, M., Osmond, H., & Newell, S. (1968). Models of alcoholism. *Quarterly Journal of Studies on Alcohol, 29*, 571-591.

Twelve Steps and Twelve Traditions. (1953). New York: A.A. World Services.

Tyrrell, I. (1997). US prohibition experiment: Myths, history, and implications. *Addiction, 92*(11), 1405-1409.

Wegscheider, S. (1981). *Another Chance: Hope and Health for the Alcoholic Family*. Palo Alto: Science and Behavior Books.

Wegscheider-Cruse, S. (1985). *Choice-making for co-dependents, adult children and spirituality seekers*. Pompano Beach, Florida: Health Communications.

Whyte, W. L. (1998). *Slaying the Dragon: The History of Addiction Treatment and Recovery in America*: Chestnut Health Systems.

Woititz, J. G. (1985). *Struggle for Intimacy*. Pompano Beach: Health Communications.

4

Exiting Public Health: Expansion of the Biomedical Model in Alcohol Research

"I am disturbed that, at a time when the committee is proceeding with thoughtful work on the reauthorization of the substance abuse block grant and the mental health block grant, the administration is suddenly and urgently proposing a major reorganization of the Alcohol, Drug Abuse and Mental Health Administration. I have yet to hear a reasonable explanation for this reorganization, and I am concerned about the financial costs and the effects on research and services. Such a dramatic change in philosophy and structure needs to be well supported by close analyses of the current problems at ADAMHA, and how the proposed changes will benefit these."
(Representative John Dingell, Chair, Committee on Energy and Commerce, Hearings on H.R. 2311 "ADAHMA Reauthorization," June 20, 1991, p 3)

Introduction

This chapter will first discuss the origins and development of the National Institute on Alcohol Abuse and Alcoholism (NIAAA) since its inception in 1970 and will then trace the forces that led to the movement of NIAAA as well as the National Institute on Drug Abuse (NIDA) and the National Institute of Mental Health (NIMH) from the U.S. Public Health Service (USPHS), and more specifically the Alcohol, Drug and Mental Health Administration (ADAMHA) where it was located from 1973-1992, to its current placement under the National Institutes of Health (NIH). This shift in location, more than anything else, symbolized and resulted in a change in NIAAA's ideology from an all encompassing biopsychosocial approach to alcohol issues to a major emphasis on viewing alcohol problems as individualistic and biomedical.

In the context of organization theory, NIAAA's move to NIH represents a change in environment that encompasses several dimensions. These domains include the extent to which the environment is stable or dynamic, simple or complex, integrated to diversified market diversity, and munificent to hostile (Mintzberg, 1979). Hasenfeld (1983) further defines the environment of an organization in two ways: the general environment and the task environment. He argues that the former focuses on economic, demographic, cultural, political-legal and technological that rarely can be affected by a specific organization. However, the task environment "...refers to a specific set of organizations and groups with which the organization exchanges resources and services and with whom it establishes specific modes of interactions" (p 51). The move of NIAAA's research mission to NIH represents a change in the task environment requiring new interorganizational arrangements to fulfill their stated goals. As will be discussed below, this shift in environment also by necessity was to include new relationships with other agencies within the NIH framework as well as with NIH itself. How NIAAA would fit into this environment is of concern to many of those testifying at the Senate and House hearings.

This chapter provides the context for chapter 5 that will illustrate in greater detail how this change in organizational location and environment within the federal government also contributed to the crystallization of the agency's ideology, goals and purpose that has had profound effects on the alcohol field in general, and on alcohol research specifically.

Brief History of the Development of NIAAA

On December 31, 1970, Public Law 91-616 known as the "Comprehensive Alcohol Abuse and Alcoholism Prevention, Treatment, and Rehabilitation Act of 1970" was signed into law by President Nixon. It had been passed unanimously by the Senate and the House of Representatives and was seen as an extremely important piece of legislation. P.L. 91-616, also known as the Hughes Act, after Senator Howard Hughes who introduced the legislation, established NIAAA as its own National Institute charged with the "...authority to develop and conduct comprehensive health, education, training, research, and planning programs for the prevention and treatment of alcohol abuse and alcoholism" (P.L. 91-616, p 1). Elliot L. Richardson, then secretary of health, education, and welfare, stated in his foreword to *Alcohol & Health, First Special Report to the U.S. Congress* (NIAAA, 1971) that the establishment of NIAAA filled the need to provide "...visibility to the depth and scope of this public health problem" (p. v). Morris E. Chafetz, M.D., served as the first Director of NIAAA until 1975.

Under this legislation several important provisions were mandated including the following: alcohol abusers and alcoholics were to be admitted for alcoholism treatment at public or private hospitals that received federal funds; federal civilian employees were to have access to alcoholism programs; federal funds were to be provided to states to encourage better programs in prevention, treatment and rehabilitation of alcohol abusers and alcoholics; all alcohol treatment records were to be confidential; the rights of recovered alcoholics in terms of hiring and firing practices in nonsecurity positions were to be protected; mechanisms were to be established for grants and contracts to promote projects for treatment and prevention services; and, developed a National Advisory Council on Alcohol Abuse and Alcoholism which reported to the secretary of health education and welfare (HEW) on a wide range issues related to alcohol abuse and alcoholism (Hewitt, 1995). Interestingly, in her article that chronicles the development of NIAAA, Hewitt (1995) points out that NIAAA did not originally have a research mandate. This was not accomplished until the passage of P.L. 94-371 seven years later.

Although NIAAA was created as its own separate agency, it was placed under the umbrella of the National Institute of Mental Health, which in turn reported directly to the Health Services and Mental Health Administration (HSMHA). Importantly, HSMHA was a part of the U.S. Public Health Service (PHS), a key agency directly under the Department of Health, Education and Welfare. From a structural standpoint, this placement of NIAAA defined alcohol abuse and alcoholism as a particular type of health and mental health problem with a major agency emphasis on clinical identification and clinical solutions. Yet, at the same time, both NIMH and NIAAA were under the USPHS, which in turn allowed and encouraged a broader interpretation of alcohol problems. It was not until the passage of P.L. 93-282 in 1974, as NIAAA was approaching its three-year authorization period, that NIAAA structurally co-existed as an equal with NIMH and the NIDA with all three agencies were placed within the newly formed Alcohol, Drug Abuse and Mental Health Administration (ADAMHA). A very important part of the 1974 legislation was that it stipulated that ADAMHA replaced HSMHA and remain under the USPHS thus implying explicitly and implicitly that alcohol abuse, alcoholism, drug abuse, and mental health issues were public health concerns as opposed to purely clinical, medical problems.

This broad focus was further supported by the passage of the Comprehensive Alcohol Abuse and Alcoholism Prevention, Treatment and Rehabilitation Act amendments of 1976 (P.L. 94-371) which augmented the mission of NIAAA to include "...behavioral and biomedical etiology of the social and economic consequences of alcohol abuse and alcoholism" (NIH Almanac). From the perspective of social scientists and behaviorists in the alcohol field, these amendments provided the necessary support for including larger issues that contributed to the social as well as individual factors associated with alcohol abuse and alcoholism.

This blend of psychosocial with biomedical perspectives is evident in the reflections of the early Directors of NIAAA (Morris Chafetz, MD, 1971-1975; Ernest Noble, MD, 1976-1978; Loran Archer (Acting Director), 1978-1978, 1981-1982, 1986) published in a 1995 issue of *Alcohol Health & Research World* commemorating NIAAA's twenty-fifth anniversary. While these early directors tended to focus more on alcoholism per se, they stressed the need to reduce the stigma associated with alcoholism by viewing it as a disease that can be treated. The role of the environment was also seen as important. For example, when asked about areas in which NIAAA has made important contributions since its inception, Dr. Noble replied "I think the public health model was one of the most important ones, and there were successful spin-offs, for example in terms of drinking and driving" (NIAAA, 1995, p. 20). In response to the question of the Institute's recent focus on molecular biology and alcoholism, Loran Archer replied "...at the moment, we are going through a period where knowledge about genes involved in alcoholism and other disease is developing very rapidly. The danger is that we get so preoccupied with the genetics that we assume genes are the total cause of alcoholism. We have to remember that genes are only one aspect of alcoholism and that the environment also plays an important role" (p 21).

It is important to remember that NIAAA was founded at a time that per capita consumption of alcohol was rising in the U.S. In 1970, average annual per capita ethanol consumption based on the population aged fourteen and older was 2.52 gallons. At that time it was the highest recorded consumption level since pre-Prohibition (1911-1915) when per capita consumption was very slightly higher at 2.56 gallons per year. Consumption levels rose dramatically throughout the 1970s to a record high of 2.76 in 1981 (USDHHS, 2004). Along with this rise of annual per capita consumption was a perceived rise in "alcoholism" within the U.S. population. Data provided by national alcohol surveys (Cahalan, Cisin & Crossley, 1969) were used to extrapolate "problem rates" and served as the empirical basis for the creation and development of an expanded mission of NIAAA to combat and prevent not only the disease termed alcoholism, but also the public health consequences of a wide range of alcohol problems. While the expanded mission was important, Cahalan (1987) adds that the "problem rates" taken from national surveys were translated to rates of alcoholism to the benefit of NIAAA. The figure of 9 million alcoholics in the population from among the more than 95 million drinkers in the nation in 1967, which was used by NIAAA in 1970, was derived from fairly arbitrary cutpoints on survey items (Cahalan, 1987).

From the Public Health Service to the
National Institutes of Health

NIAAA, along with NIDA and NIMH, remained a part of ADAMHA until the early 1990s. Under the USPHS, these agencies housed not only their research activities but also services and programs aimed at treatment and prevention within each of these three realms: alcohol, drugs and mental health. In 1991, briefly after Dr. Louis W. Sullivan, then the secretary of health and human services made an announcement, Congress introduced two bills (House and Senate) calling for a restructuring of the existing ADAMHA structure. Senator Ted Kennedy introduced the bill, co-sponsored by Senator Orrin Hatch, in the Senate entitled *The ADAMHA Reorganizaiton Act of 1991 and Related Matters (S. 1306)*. This bill was "to amend Title V of the Public Health Service Act to revise and extend certain programs, to restructure the Alcohol, Drug Abuse and Mental Health Administration, and for other purposes." A parallel bill was introduced in the House entitled *ADAMHA Reauthorization (H. 361)*. This restructuring of ADAMHA would place it as an umbrella agency (under the new name of Alcohol, Drug Abuse and Mental Health Services Administration—later renamed as the Substance Abuse and Mental Health Services Administration—SAMHSA) for three separate centers that would focus on treatment and prevention services: the Center for Substance Abuse Treatment (CSAT), the Center for Substance Abuse Prevention (CSAP), and the Center for Mental Health Services (CMHS). It also called for the movement of the research functions of NIAAA, NIDA and NIMH to the National Institutes of Health (NIH) where they would be devoid of their current service components. While several other provisions of these bills relate directly to the reorganization of ADAMHA, they primarily were aimed to separate research and services within NIAAA, NIDA and NIMH; to combine alcohol and drug treatment and prevention administratively in the form of two Centers; and, to create a separate Center that provided support for mental health services (Midanik, 2004).

For NIAAA and NIDA, this proposed reorganization represented a major change both administratively and politically. However, for NIMH it was different. Originally, NIMH was an agency within NIH when it was created in 1949, and it was designed "...to coordinate the research, training, and service activities authorized by the National Mental Health Act of 1946..." (Editorial, 1991a, p. 861). In 1967, NIMH was transferred out of NIH and was placed under the Health Services and Mental Health Administration (HSMHA) until 1973, when it joined NIDA and NIAAA under ADAMHA (after briefly being back under NIH for less than one year when HSMHA was disbanded in 1973). An attempt was made in 1987 to bring NIMH back to NIH which was strongly supported by The National Alliance for the Mentally Ill (NAMI), a mental health consumer advocacy group. Similar to the argument made in the alcohol field, NAMI felt that the movement to NIH would provide credence to the idea that mental illness was a disease. However, this earlier attempt to return NIMH to NIH was not approved by Otis R. Bowen, M.D., the secretary of health and human services at that time.

Under the leadership of Representative Henry A. Waxman, hearings were held before the Subcommittee on Health and the Environment of the Committee on Energy and Commerce on June 20, 1991. Senator Ted Kennedy, chair of the Senate Committee on Labor and Human Relations chaired the Senate, Hearings on June 25, 1991. Materials given and provided for both hearings and the information presented in the 1991 Institute of Medicine Report entitled *Research and Service Programs in the PHS: Challenges in Organization,* provided the reasoning behind this proposed reorganization of ADAMHA. It also illustrates the potential problems of NIAAA's move, particularly for behavioral and social sciences, to the NIH (Midanik, 2004).

The hearing before the House Subcommittee on Health and the Environment (Representative Waxman, Chair) is a 250-page document comprised of testimony and supporting materials (*ADAMHA Reauthorization (H. 361)*, 1991). Even before the hearings began, Representative John Dingell, chair of the Committee on Energy and Commerce, expressed his concern as stated in the quotation at the beginning of this chapter. Representative Waxman also made reference to his concerns regarding the proposed reorganization in his opening remarks. First, he questioned how NIAAA, NIDA and NIMH would survive or at least be noticed under NIH given "...an already crowded NIH portfolio" (p 1). Second, Representative Waxman expressed concern about whether this legislation represents a change in "the administration's historic resistance to an expanded Federal role in services, particularly mental health services" (HR p. 1).

In essence, these bills and their Hearings reflect a two-part process. The first is the restructuring of ADAMHA making its sole focus on treatment and services in the alcohol, drug and mental health fields. The second, and perhaps less understood arena, is the possible consequences of separating the research functions of these three areas from services and treatment and moving them to the National Institutes of Health. Much of the testimony in both Hearings focuses on the former—providing the reasons for the importance of separating out, and enhancing treatment and prevention.

Tension between Services and Research

Perhaps one of the most compelling questions that both Committees asked, albeit not directly, is the "Why now?" question. The specific rationale for this reorganization is best described in the Institute of Medicine (IOM) report in which the history of ADAMHA is outlined in great detail (IOM, 1991). Concerns regarding the ability of a single agency to handle effectively the service, training and research components of mental health (the so-called "three-legged stool") were expressed at NIMH's inception in 1946. Interestingly, even though NIMH was placed within NIH, its research function varied considerably from the other Institutes: "In addition to basic and clinical biomedical research, NIMH strongly supported behavioral research and some social science research" (IOM, 1991, p. 28). The IOM report goes on to explain that this was an understandable emphasis given the lack of knowledge concerning the biological etiology of mental disorders.

Several legislative events occurred in the 1950s and 1960s that shifted the focus of NIMH to service delivery—primarily treatment. Most notably The Community Mental Health Centers Act of 1963 was passed which led to public funding for treatment facilities for severe mental health problems of both children and adults. In 1966, the National Center for Prevention and Control of Alcoholism became part of NIMH as well as the Center for Studies of Narcotic Addiction and Drug Abuse. Because of the expansion of the service function of NIMH, the research community was unhappy and feared that its importance would be minimized and that its funding was being increasingly directed at service functions within the agency. By 1967, NIMH was the largest agency within the NIH, accounting for 22 percent of NIH's budget. In 1967, NIMH was moved out of NIH and became a bureau; however, by 1968 it was moved again to the Health Services and Mental Health Administration (HSMHA) whose function was to manage PHS service delivery programs (IOM, 1991).

These changes in the 1960s did not halt the controversies over the mission of NIMH and its placement as a federal agency. The Comprehensive Alcohol Abuse and Alcoholism Prevention, Treatment and Rehabilitation Act of 1970 and the Drug Abuse and Treatment Act of 1972 eventually established two additional Institutes that structurally were no longer located under NIMH. In 1973 after a major reorganization of the USPHS, NIMH went back to NIH although quite briefly (IOM, 1991).

Concerned voices from the research community with regard to NIMH's triple mission and its lack of emphasis on research resulted in the formulation of a Task Force and a subsequent report ("The Gardner Report") that recommended five options for how NIMH (and ultimately NIAAA and NIDA) could best be organizationally situated. The secretary of health, education and welfare decided to develop ADAMHA, which would house all three institutes and be located under PHS. Interestingly, the Task Force expressed the view that the drug and alcohol areas should eventually be combined but always remain separate from the mental health function (IOM, 1991).

The struggles outlined above between the service and research functions of each agency continued to occur. In 1981, under the Reagan administration, the Alcohol, Drug Abuse and Mental Health Block Grant Program was created which encompassed both categorical and formula grant programs to the states. Thus, more emphasis was placed on the agency's research functions—much to the dismay of service advocates. However, within five years the Anti-Drug Abuse Act of 1986 was passed that provided more funds for demonstration programs in community agencies and for preventive services. This was followed by additional legislation that infused more money into ADAMHA for service programs (IOM, 1991).

These continuing organizational changes illustrate the ongoing struggle between the service and research functions of each agency within ADAMHA. The National Alliance for the Mentally Ill (NAMI) and some researchers lobbied for NIMH to move back to NIH so that its research mission would be enhanced and thus, they hoped, the contention that mental illness is a treatable disease would be organizationally endorsed. In 1987, Lewin and Associates were commissioned to suggest strategies for how ADAMHA could be reorganized to suit the needs of its service and research missions. Like the Gardner Report of 1973, the Lewin Report offered five options based on sixty-two interviews with key informants from the various interest groups. Ultimately, the decision was made to keep ADAMHA. This decision was justified by noting that "...although destigmatization and integration of ADM activities are appropriate goals for DHHS, organizational change alone would not achieve these objectives" (IOM, 1991, p. 40).

Further legislation in 1988 that resulted in the creation of the Office for Substance Abuse Prevention and the Office for Treatment Improvement also called for an additional study by the Institute of Medicine of the implications of focusing ADAMHA's efforts on service-oriented activities and separating its research functions. The results of that 1991 IOM study are discussed later in this chapter.

Much of the conflict between the service and research communities is voiced in the Congressional Hearings. The argument was made that as funding for categorical service programs ended and block grants emerged, more emphasis was placed on research and less on treatment and prevention (Hearing HR, 1991, p. 5). The general feeling that treatment services and prevention activities had somehow been slighted or seen as less important provides another theme throughout these Hearings. However, with the advent of President Reagan's and President George Bush's national drug control strategy and additional funding for prevention and treatment services, which included the creation of the Office of Substance Abuse Prevention and the Office of Treatment Improvement, the research community began to feel "slighted."

...the budget for drug abuse prevention has more than doubled since 1989. Managing
this rapidly growing service responsibility and those for alcohol abuse and mental
health has become increasingly difficulty with the competing research priorities (*The
ADAMHA Reorganization Act of 1991 and Related Matters (S. 1306,* 1991). (Testimony of Dr. James O. Mason, assistant secretary for health, p. 7)

The result of this additional drug abuse prevention funding, accord-
ing to Dr. Mason, has been a tension between the two functions
within each of these agencies resulting in significant and ongoing
administrative problems.

Progress related to basic biological, biomedical and behavioral research priorities have
been somewhat overshadowed by growth in ADM service programs and funding with
their different and competing priorities have led us....to conclude that the benefits of
separately administering the services and research programs will outweigh the advantages of leaving them together. The new ADAMHA will benefit as this reorganization
allows resources to be focused in a manner that enhances treatment and prevention
programs while supporting a strong research agenda. (Testimony of Dr. James O.
Mason, assistant secretary for health, June 25, 1991 Senate Hearing, p. 7)

In summary, the Hearings in Congress served as a venue for discussing larger political concerns related to funding and attention
focused on research versus services. It was presumed by many of
those who testified that if treatment and services were separated,
additional financial resources would be made available. In fact, some
attention at both Hearings was focused on the special treatment needs
of pregnant substance-using women. It is important to note that the
third part of the "three legged stool"—training—was mentioned mini-
mally in the Hearings.

Potential Problems of Moving to NIH

A primary reason for the movement of NIAAA, NIDA and NIMH
from ADAMHA was to transfer the research functions of these agencies to NIH and perhaps to legitimize these functions by aligning
them with the biomedical research directions of NIH. This was consistent with the declaration of President George Bush and Congress
that the 1990's were to be called the "Decade of the Brain." Twice in
his testimony before the Senate Subcommittee, Dr. Mason mentions
the increased benefits of this move for the three agencies and, to a
lesser extent, the benefits that will be derived for NIH when NIAAA,
NIDA, and NIMH joins it.

Placing the three research institutes within the NIH will provide greater support for
scientific achievement; it will increase visibility and prestige and open new avenues for
research collaboration with the other institutes at NIH. The benefits of these increased

research interactions will accrue to both the existing NIH institutes as well as the transferred institutes. (p. 7)

...I see this as a win-win for the three institutes. They will be able to enjoy the benefits of the rest of NIH in the fields of biotechnology, neuropharmacology, genetics, some of those cutting-edge sciences, while the rest of NIH will benefit from the bio-behavioral expertise of those three institutes that are coming in from ADAMHA. (p. 9)

Note that the "benefits" for the three agencies are framed by the added status of being within NIH and thus having better access to biomedical research that is seen as the foundation of NIH. This idea was also expressed by Dr. Frederick K. Goodwin, then administrator of ADAMHA, in his testimony before the House Subcommittee.

This merger fulfills a 25-year dream of mine—widely shared by the research community and patient groups—to see the reintegration of the mental and addictive disorders into the mainstream of biomedical research. I believe there is unparalleled opportunity for our three institutes as full members of NIH, which has always represented—here and abroad—government at its best and science at its best. (Testimony HR, 1991, p. 30)

Perhaps a bit skeptical of Dr. Goodwin's testimony and his apparent "change of heart," Representative Waxman reminded him of a memo Dr. Mason, then assistant secretary for Health, had prepared in April 1990 in which he stated that moving these research institutes to NIH might increase their prestige, yet "...they would represent a small piece of an already overburdened NIH, certain to receive little attention" (Testimony HR, p. 31). Dr. Goodwin countered this comment by noting that with NIH's new leadership, Dr. Mason's original thoughts about this proposed move changed.

Representative Waxman continued with these questions asking Dr. Goodwin to comment on another passage in Dr. Mason's memo, which stated that "It is difficult to imagine, for example, the alcohol and drug abuse research programs competing successfully for time, attention, and resources within the NIH milieu" (p. 32). Dr. Goodwin stated (after being asked a second time) that assurances have been secured from Secretary Sullivan that the three Institutes would remain unchanged in the transfer to NIH.

One of the most compelling questions came from Senator Durenberger who asked Dr. Mason during the Senate hearing "Do we have the right capacities in place to deal well with some of the research aspects of the behavioral sciences?" Dr. Mason's response is quite revealing with regard to his conceptualization of behavioral sciences.

What this does is it provides a new continuum. For example, you can start with the biochemistry of a single brain cell, and the continuum extends from that cell and its neurophysiology and extends to complex human behavior and its effect on health and disease, and between the basic biochemistry and the behavior of an individual, we can now begin to focus all of the tools that are available at NIH– for example, the genetics, the human genome, not only those neurosciences that have been explored and expanded by ADAMHA but those that are available at NIH. So it provides tools that would have been more difficult to apply while they were in separate agencies. (Senate Hearings, 1991, p. 12)

Note that nowhere in Dr. Mason's testimony is there reference to the importance of the environment or policies that could help buffer an individual from disease or negative behavior related to alcohol use. Perhaps Dr. Mason represents the NIH biomedical approach to disease which is individualistic by definition. Thus, instead of behavioral research encompassing everything that is not biomedical, it is clear it is meant to be part of a truncated "continuum" that does not appear to consider any environmental or policy issues.

These concerns about what might occur given the move to NIH were also voiced in several prepared statements submitted to the Senate Subcommittee. In its prepared statement, the American Psychological Association noted their concerns for the

...vitality of the full range of behavioral and social sciences research: There is concern that behavioral and social sciences research, particularly applied research, will not fare well when housed with the biomedical institutes at NIH. While more biomedically oriented research now being conducted at the ADAMHA institutes is likely to fare well at NIH, it is not clear that a similar level of support exists for the broader portfolio of behavioral and social sciences research." (Senate Hearing, p. 71)

These concerns are also expressed by the American Psychological Society (APS) that supported the Senate bill but took exception to several issues concerning NIMH. One area is Section 305 of the Bill which called for a uniform definition of "serious mental illness" to be developed. APS responded strongly to this emphasis on serious mental illness to the exclusion of other types of mental illness:

We see this as more than simply a request for a definition to be developed....it is a political move that will trivialize such things as children's behavioral disorders, panic attacks, and the host of debilitating mental and emotional disorders that are the result of child abuse, addiction, community violence, or severe cognitive disabilities–disorders that disrupt individual lives in very real ways. We see the promotion of "serious mental illness" over other disorders as divisive for the field at a time when we need to unite behind a re-constituted NIMH. (Senate Hearings, p. 73)

Even more strongly, Alan Kraut, Executive Director of APS, makes the argument that the relationship between health and behavior is increasingly seen as being outside of HHS' mission. He provides as an example Secretary Louis Sullivan's call for a new "culture of character" to handle the myriad of social problems in which there is a behavioral component. Kraut contends that Secretary Sullivan's statement indicates that he does not see:

> …issues of behavior and lifestyle choices as being appropriate topics for science, but rather as issues of morality and values. The Secretary talks in terms of "unwise choices" in lifestyle; but to the person in the throes of physical or psychological addictions or other behavior-based conditions, controlling undesired behavior often is no longer a matter of free will or individual choice. The message being conveyed is that those individuals require nothing but the stimulus of their own problems in order to solve them. (Senate Hearing, p. 74)

Kraut thus soundly critiqued the medical model approach to social problems and its limitations and made a strong argument that behavioral research is complex and varies considerably from traditional medical strategies. He goes on to state:

> I regret to say that this [Sullivan's comments] is a perfect illustration of the traditional medical approach to health and behavior–if the medical model fails to apply, if there isn't a magic bullet in the form of a vaccination or a pill, then each individual is pretty much on his or her own to work it out. In fact, while lifestyle changes and increased individual responsibility are desirable goals, reaching these goals requires knowledge about the origins of the undesired behavior, what change are needed, and how to make those changes. (Senate Hearings, p. 74)

This skepticism is also clearly expressed by the prepared statement of the American Society of Addiction Medicine (ASAM). Under the section entitled "Alcohol Prevention Research," ASAM is critical of NIAAA's nearly exclusive focus on biomedical research even within ADAMHA. The ASAM statement astutely points out that some areas of alcohol research clearly would not fit within the realm of "behavioral" research in the context of the biomedical ideological purview of NIH. They cite research areas such as drinking and driving crashes or addiction prevention. They argue that a "different type of research" is needed to address these problems: research that would stem from econometrics, behavioral, and social sciences.

This concern for the future of behavioral and social science research is expressed most concisely in the Senate hearings by Bob Galea, a former chronic drug abuser, drug-free for twenty-four years, and someone who established a drug treatment and research center in Massachusetts. While supporting the reorganization of ADAMHA he notes:

It is going to be important in this legislation to assure that behavioral research survives and not be dominated by the rigorous biomedical research approach of NIH (Senate Hearing, p. 53)

In summary, based on the material in the Senate and House Hearings, there seems to be both enthusiasm and cautious concern for the potential consequences of moving NIAAA, NIDA and NIMH to NIH. Part of the issue appears to lie in conflicting understanding of the definition of "behavioral research." In the language of NIH, behavioral research appears to be anything *other* than biomedical research. As such, this definition is very broad and, to some extent, also quite vague. While under this definition there would be no question that traditional treatment evaluation studies, for example, could be defined as behavioral research, other areas are far more questionable. Would population studies that examine the effects of taxation on alcohol use among underage drinkers be considered behavioral research? Would assessing the alcohol beverage industry's targeting of minority youth in their advertising also fall into that category? In other words, how broad is behavioral research? As shown in statements from both hearings, some individuals did extend this concept and included "social sciences," but in general, most seemed to be content to use the dichotomous categorization dictated by NIH.

The Institute of Medicine (IOM) 1991 Report

Originally, when Congress commissioned the IOM to conduct a study that assessed the possible reorganization of ADAMHA, the expectation was that it would be completed prior to the Congressional Hearings. In both the Senate and the House Hearings, references are made to the IOM report and whether or not endorsement of this proposed reorganization is premature until the IOM report is completed. The concern expressed by both subcommittees is evident. In the Senate:

The Chairman. A point has been made by some of our colleagues that we ought to wait until the Institute of Medicine study is completed and that we are premature in terms of developing this approach. What is your reaction? I know you have had access to and have spoke with those in the Institute of Medicine, and I'm just wondering how you think we should respond to that recommendation or suggestion that we ought to wait.

Dr. Mason. I am sure that you and the department have been very interested and following along with the Institute of Medicine study. We had anticipated that their report would be out by the 15th of May. There were some delays that occurred at the Institute of Medicine level, and for that reason we were unable to wait.

I have been in discussions with both the chairman of the committee that is studying this issue as well as Dr. Sam Thier, and I have been told that what we are proposing is not inconsistent with what will be the final report.

We found and thought that at a time when Congress was looking at the reauthorization of ADAMHA that this was the time to move forward, and we feel when we are finished and the Institute of Medicine is finished that it will all come together and fit well. (S.B 1306 Hearing, p. 8)

In the House:

Mr. Waxman. Now we commissioned a study by the National Academy of Sciences to look at this issue and they have spent–they have contracted costs of more than $750,000. Why did the Department move forward on the reorganization before we even had that study addressing the appropriateness of such a plan?

Mr. Goodwin. I think the Department, by the way, has had ongoing discussions with IOM and has had interactions with the IOM committee, and have obtained some sense of the types of findings and recommendations that will be forthcoming later in the summer. We also anticipate the legislative review of this reorganization plan will allow time for the incorporation of the IOM findings.

It was organizationally supposed to come out in May, and there was a delay in that, which is part of the reason for what seems to be a lack of synchrony. (HR 102-44, p. 33)

The IOM report itself is based on several sources of data. It included: 1) case studies and background papers; 2) a review of previous studies conducted within NIH, ADAMHA and other agencies; 3) a review of policy papers; and, 4) an analysis of grants, service development, programs and projects between the years of 1975-1989 within specific case study areas, e.g., Alzheimer's Disease, Substance-Abusing Pregnant Women, Dopamine Research, Schizophrenia and Parkinson's Disease. As stated in the preface to the IOM report, over 150 interviews were conducted with a wide range of interest groups across agencies and constituencies.

In the final product, the IOM does not wholeheartedly support the proposed changes in ADAMHA, perhaps as originally hoped for by its supporters. Rather, after considering previous reorganizations and their original intentions and actual results, the report is quite cautious. One of its two recommendations for Congress is put quite simply: "If reorganization of current agency structure is considered, it should be justified on policy grounds" (IOM, 1991, p. 1). While supporters of this reorganization can and do make the argument that these shifts are policy based, the IOM elucidates this point further by arguing that previous efforts at administrative changes have operated on the assumption that these changes will make a difference in service delivery. "The experience of these years, however, suggests that such a relationship may be an act of faith rather than demonstrated by evidence" (IOM, 1991, p. 106).

Another recommendation of the IOM report for the secretary of health and human services was that "below the agency level, research and service programs should be administered and conducted by separate institutes or offices" (IOM, 1991, p. 1). Further, the report does not support the idea that moving NIAAA, NIMH and NIDA to NIH will have any affect on resources allocated to these agencies (Editorial, 1991b).

Thus, we can see that the expected overwhelming support for the separation of services and research within the realm of alcohol, drug use and mental health was not provided by the IOM report. Instead, the IOM report pointed to the need to handle more basic issues within ADAMHA, such as its mission since the implementation of the Block Grant Program of the 1980s, which they deemed as more timely and important. There appears to have been little attention paid to these prudent IOM recommendations.

Conclusions

Despite some of the hesitancy and caution expressed in the Hearings and the mixed results of the IOM study, NIAAA, NIDA, and NIMH did obtain Congressional approval to move their research functions to NIH and to leave treatment and prevention services at ADAMHA. To handle these functions, three new centers were developed eventually under a new organization called the *Substance Abuse and Mental Health Services Administration* (SAMSHA). These three centers are: The Center for Substance Abuse Treatment, The Center for Substance Abuse Prevention, and The Center for Mental Health Services. Unlike Redman's (1973) description of the creation of the National Health Service Corps, the "dance of legislation" around moving NIAAA to NIH was relatively quiet and brief. Perhaps this was due to an increasing interest in biomedical research already included in the goals of NIAAA. Thus, this change in institutional auspices may have been seen as a "natural evolution" given the priorities of the environment.

Yet, NIAAA's move from ADAMHA to NIH signaled many things in the alcohol field; most importantly, it shifted NIAAA's primary focus to "alcoholism" as opposed to the multi-faceted concern with "alcohol abuse." This in turn resulted in an even stronger dominance of biomedical models wherein the etiology, detection, prevention and treatment of alcoholism became the major concern.

Chapter 5 examines in detail some of the consequences of this organizational and ideological shift in terms of the attention placed on individualized explanations of etiology, diagnoses and cure. Moreover, it demonstrates how this focus became legitimized within the larger context of the National Institutes of Health.

References

ADAMHA Reauthorization (H. 361), House of Representatives (1991).

The ADAMHA Reorganization Act of 1991 and Related Matters (S. 1306). Senate (1991).

Cahalan, D. (1987). *Understanding America's Drinking Problem. How to Combat the Hazards of Alcohol*. San Francisco, CA: Jossey-Bass.

Cahalan, D., Cisin, I., & Crossley, H. (1969). *American Drinking Practices: A National Study of Drinking Behavior and Attitudes*. New Brunswick, NJ: Rutgers Center of Alcohol Studies.

Editorial. (1991a). HHS Plan would shift alcohol, drug and mental health research to National Institutes of Health. *Hospital and Community Psychiatry, August*, 860-861.

Editorial. (1991b). Institute of Medicine Study Offers Little Support for Transfer of ADAMHA Research to NIH. *Hospital and Community Psychiatry, October*, 1077.

Hasenfeld, Y. (1983). *Human Service Organizations*. Englewood Cliffs, NJ: Prentice-Hall, Inc.

Hewitt, B. G. (1995). The creation of the National Institute on Alcohol Abuse and Alcoholism: Responding to America's alcohol problem. *Alcohol Health & Research World, 19*(1), 12-16.

IOM. (1991). *Research and Service Programs in the PHS: Challenges in Organization*. Washington, DC: National Academy Press.

Midanik, L.T. (2004) Biomedicalization and alcohol studies: implications for policy. *Journal of Public Health Policy, 25*(2), 211-228.

Mintzberg, H. (1979). *The Structuring of Organizations. A Synthesis of the Research*. Englewood Cliffs, NJ: Prentice-Hall, Inc.

NIAAA. (1971). *Alcohol & Health, First Special Report to the U.S. Congress*.

NIAAA. (1995). Reflections: NIAAA's Directors Look Back on 25 Years. *Alcohol Health & Research World, 19*(1), 17-59.

Redman, E. (1973). *The Dance of Legislation*. New York: Simon and Schuster.

USDHHS. (2004). *U.S. Apparent Consumption of Alcoholic Beverages*. Bethesda, MD: Public Health Service, National Institutes of Health, National Institute on Alcohol Abuse and Alcoholism, NIH Publication No. 04-5563, June.

5

Manifestations of the New Ideology

"I begin with the premise that behavior is an incredibly important element in medicine," he said, "People's habits, their willingness to quit smoking, their willingness to take steps to avoid transmission of H.I.V., are all behavioral questions.

"But what I'm looking for are new ideas, real discoveries. When I read about genetics, I see breakthroughs every day. And while I'm trying to learn more about behavioral science, I must say that I don't feel I get tremendous intellectual stimulation from most of the things I read." (Dr. Harold Varmus, then newly appointed Director of NIH, New York Times, November 23, 1993).

Introduction

In 1992, when NIAAA made the move to NIH along with NIDA and NIMH, there was some speculation about the influence NIH might have on these agencies. While NIH may include health in its title, its primary focus has been on disease and pathology. Skepticism concerning whether or not these agencies could maintain a balanced focus on prevention research, treatment research, and basic research was expressed during the Hearings as discussed in chapter 4. There is evidence, even early on, NIH did, in fact, have a pronounced effect on how NIAAA was presented in terms of its mission and priorities.

In the same quarterly report of *Alcohol Health & Research World* in 1995 commemorating NIAAA's twenty-fifth anniversary, the lead article, written by Enoch Gordis, M.D., then Director of NIAAA, was entitled "The National Institute on Alcohol Abuse and Alcoholism. Past Accomplishments and Future Goals." In several ways, this article clearly reflects the dominance of the biomedical model and provides the backdrop for the focus of this chapter. The article continually focuses on "alcoholism" and the biomedical advances taking place at NIAAA (Gordis, 1995). While Dr. Gordis does concede that "...many alcohol related problems result from misuse of alcohol by persons who are not alcoholic" (p. 5), he nonetheless goes on to stress the importance of focusing on alcoholism (the disease concept) because it has apparently "...sharpened the focus of alcohol research..." (p. 5). The rest of the article primarily discusses genetics, molecular genetic studies, and medical effects. Space is provided for environmental factors (mostly in relation to its interaction with genetics), with no mention of cultural determinants of alcohol use. Prevention and treatment are only briefly discussed at the end of the article, giving the impression that they are of lesser importance to the goals of NIAAA. The explicit message is that biomedical research is the priority at NIAAA above all other research.

This message is consistent with a personal interview with Dr. Gordis also featured in this issue of *Alcohol Health & Research World* (1995). When asked by the interviewer "What, in your opinion, were the primary achievements of NIAAA during the past 25 years?" Dr. Gordis replied:

> One could list numerous important scientific insights concerning treatment, genetics, the development of animal models, or the importance of neuroscience. But I think what counts most is that NIAAA showed that research can make an important contribution to solving the alcohol problem. Alcoholism treatment can be approached with the same scientific rationale and style as other areas of medicine. It has stopped being a vaguely formulated problem and instead gradually has been brought into the mainstream of medical science. The individual findings are important, but the general recognition that alcoholism can be studied with the most contemporary tools of science is the contribution that stands out. (p 26)

When one assesses the list of NIAAA achievements discussed above, it is clear that policies focusing on environmental strategies, such as limiting access and availability to alcohol, are not considered an important part of NIAAA's mission. Dr. Gordis' emphasis on the "ism" negates, or at the very least minimizes, the importance of looking at the broader field of alcohol problems from a public health perspective.

One only needs look on NIAAA's webpage (www.niaaa.nih.gov) to know that NIAAA has moved more towards a biomedicalization approach. In its statement of purpose, it states clearly:

> The National Institute on Alcohol Abuse and Alcoholism (NIAAA) supports and conducts biomedical and behavioral research on the causes, consequences, treatment, and prevention of alcoholism and alcohol-related problems...(NIAAA, 2002)

It is not clear if this dichotomy of biomedical and behavioral research mentioned in this statement completely omits social, cultural and/or environmental factors that affect "alcoholism and alcohol-related problems" or if these areas are included under the umbrella of "behavioral" research. In either case, it is clear that NIAAA's priority is biomedical research.

The purpose of this chapter is to illustrate the growing biomedicalization that has occurred in the alcohol field prior to and enhanced by NIAAA's movement to NIH. This shift in place and in ideology will be traced through several public documents published by NIAAA, beginning with its most recent strategic plan and including the last ten *Alcohol & Health Reports to Congress* published by NIAAA over the period of 1971-2000. In addition, trends in funding of research by NIAAA will be examined between 1990 and 2003 to assess the relative importance NIAAA places on biomedical research versus non-biomedical research (termed "behavioral research" by NIH).

Strategic Plan 2001-2005

In Midanik's (2004) article, the NIAAA's Strategic Plan for 2001-2005 is discussed in terms of its seven goals and is compared to NIDA's strategic plan written approximately at the same time. While NIAAA's current plan was written in response to "...to the requirements of the Government Performance and Results Act (GPRA) ...passed in 1993, requires NIAAA to plan and measure performance in new ways" (NIAAA, 2000, p. 1) the only other NIAAA Strategic Plan was written in September 1986 (personal communication with Brenda Hewitt, special assistant to the director, October, 2000). This plan, entitled "Toward a National Plan to Combat Alcohol Abuse and Alcoholism. A Report to the United States Congress," was written as a report to Congress to define the extent of alcohol abuse and alcoholism, to develop objectives and strategies for combating alcohol abuse and alcoholism, to justify the important role of research in this area, and to clarify the role of the federal government, states, local governments and the private sector in providing programs, services and research to combat alcohol abuse and alcoholism.

Within this first strategic plan, the following six objectives are identified: (1) Reduce prevalence of alcohol abuse and alcoholism among adults aged eighteen and older by 29 percent by 1995; (2) Reduce prevalence of acute drinking-related problems among youth aged fourteen to seventeen years by 11 percent by 1995; (3) Reduce the rate of deaths from motor vehicle accidents involving drivers with blood alcohol levels of .10 percent or more by 19 percent by 1995; (4) Reduce the rate of non-vehicular alcohol-related traumatic injuries by 29 percent by 1995; (5) Reduce the death rate from liver cirrhosis by 11 percent by 1995; and (6) Reduce the incidence of fetal alcohol syndrome by 25 percent by 1995 (NIAAA, 1986, pp. 12-13). In this report, the rationale for the choice of these objectives and their proposed percentage reduction is based on lowering overall financial costs to society. Note that the objectives are framed primarily in terms of reduced morbidity, mortality, and negative consequences associated with excessive use of alcohol. Very little emphasis is on "alcoholism" per se.

Written approximately fifteen years later, the 2001-2005 Strategic Plan is a comprehensive and important document that designates what the current and future directions of NIAAA will be. It states clearly that its purpose is to insure that "...its resources are invested wisely" as well as to clarify which research areas will be strongly supported and "moved ahead on an accelerated track." Thus, it is clear that this document provides important insight as to what the agency perceives is its most important mission and what areas need to be investigated to achieve their goals. What is most remarkable about the Strategic Plan document is its emphasis on biomedical achievements and its omission of social science as an important area in need of research support. This is problematic given the claim in the text of the Strategic Plan that NIAAA supports "...90 percent of all alcohol research conducted in the US" (p. 2). This means that one agency does have the ability to influence the field in determining which research areas will be encouraged and supported and which ones will not.

There are seven goals included in the 2001-2005 Strategic Plan that represent present and future priorities. These seven goals are listed below.

Goal 1. Identify genes that are involved in alcohol-associated disorders.

Goal 2. Identify mechanisms associated with neuroadaptation at multiple levels of analysis (molecular, cellular, neural circuits, and behavior).

Goal 3. Identify additional science-based preventive interventions (e.g., drinking during pregnancy and college-age drinking).

Goal 4. Further delineate biological mechanisms involved in the biomedical consequences associated with excessive alcohol consumption.

Goal 5. Discover new medications that will diminish craving for alcohol, reduce the likelihood of post-treatment relapse, and accelerate recovery of alcohol-damaged organs.

Goal 6. Advance knowledge of the influence of environment on expression of genes involved in alcohol-associated behavior, including the vulnerable adolescent years and in special populations.

Goal 7. Further elucidate the relationships between alcohol and violence.

Given that these goals reflect the current and future focus of NIAAA, it is important to reflect on the content areas they include and do not include. Midanik (2004) argues that five of the seven goals have a biomedical focus—the exception being Goal 3 (Prevention) and Goal 7 (Violence). Treatment is mention only in the context of the development of medications to lessen craving (Goal 5), and environmental issues are discussed only in reference to the expression of genes (Goal 6). Not only do five of these goals reflect the dominance of biomedical issues, the order of the goals also reflects their importance. The first two goals are "purely" biomedical; with the exception of prevention, goals that mentioned the environment or treatment are further down on the list. While having the goal of elucidating the relationships between alcohol and violence is commendable and extremely important, it is last on the list and likely not seen as important as the other more biomedically-focused goals (Midanik, 2004).

Other Strategic Plans

In sharp contrast to NIAAA's most recent strategic plan, NIDA's five-year strategic plan covering the same time period has only one goal: "To significantly reduce drug abuse and addiction and their behavioral, health, and social consequences" (NIDA, 2000). Unlike NIAAA, the introduction to NIDA's Strategic Plan does not dichotomize biomedical and behavior research. Rather it includes social research as a separate category and recognizes that it differs from other levels of research. Treatment and prevention efforts are also highlighted in NIDA's strategic plan. Included with NIDA's one main goal are three strategies for its achievement. Strategy 1 is "Give communities science-based tools to prevent drug abuse and addiction"; Strategy 2 is "Develop and distribute tools to improve the quality of drug abuse treatment nationwide"; and, Strategy 3 is "Educate the public about drug abuse and addiction." Supporting each of these three strategies is the effort to increase understanding on the nature of addiction through basic and clinical research. While the goals of both agencies may, overall, be fairly similar, their approaches are quite different. Genetics and Neurobiology are included as part of both Strategy 1 and Strategy 2, but they are not emphasized as the most important components. Overall, NIDA's Strategic Plan provides a strong balance between treatment, prevention, and biomedical/genetic issues.

As mentioned above, only one other NIAAA Strategic Plan could be located, and it was written conducted in 1986. Its title, "Toward a National Plan to Combat Alcohol Abuse and Alcoholism," suggests a united approach to prevent and eradicate the consequences of alcohol abuse and alcoholism. The document presents a very balanced focus and applies the issues of defining the problem, prevention strategies, interventions and treatment needs to both alcohol-related problems and alcoholism. While biomedical research is clearly discussed as a research area that warrants further support, it is framed within the context of prevention/intervention and treatment. This is also true for genetic research. Further, the document places research relating to organ pathologies in the context of the costs of alcohol-related illnesses. Thus, the 1986 plan differs dramatically from the current plan written fifteen years later in terms of NIAAA's vision of what it represents and the directions it plans to go. It can be argued that in 1986, biomedical research in the alcohol field was less advanced and that newer biotechnical tools have been developed since that time making its importance more salient. While this may be the case and that biomedical and biotechnical advances have made specific types of research possible, it is critical to consider the importance of integration and balance. The 1986 Strategic Plan combines and synthesizes the research areas so that the one area does not overshadow another. It emphasizes the blending and interdependent relationships between and among areas.

Alcohol and Health: Reports to Congress

Coverage of Content

From 1971 until 2000, NIAAA produced ten comprehensive reports to Congress entitled *Alcohol & Health* designed to describe the "state of the field" and to showcase the progress in the research arena almost exclusively funded by NIAAA. Approximately every three years between 1971 and 2000, a task force was appointed to identify which would be included and how these areas would be organized and addressed. In addition, the task force solicited experts in the field who were willing to write comprehensive overviews of these areas. Typically, these authors are acknowledged in a group and not by specific contribution; thus, NIAAA reserved the right to edit these chapters. In many ways, *Alcohol & Health* is NIAAA's opportunity to present itself to Congress, and to the nation in general, and clarify what it has accomplished and what more needs to be done to combat the problems of alcohol abuse and alcoholism. Over the more than three decades of NIAAA's existence, these ten reports have reflected the ways in which NIAAA wants to be viewed by its various constituencies. By assessing what areas are included in these reports, the relative attention given to specific areas, and the placement of these topics in these reports, one can get a good idea of how NIAAA constructs its "presentation of self" both within its umbrella agency (originally ADAMHA, now NIH) and the federal government, as well as in relation to other federal agencies in general.

To determine coverage of substantive areas, a content analysis was conducted on the ten *Alcohol & Health* reports that assessed coverage of five main content areas as measured by the percentage of pages within each report devoted to a specific topic (Midanik, 2004). Two independent researchers coded each of the ten reports, and there was a high agreement (90 percent) between them. A detailed description of the coding for each of the ten volumes is included in Appendix A. When disagreements occurred, a consensus was reached, usually by double coding (or triple coding) pages of chapters devoted to more than one topic area such as fetal alcohol syndrome or fetal alcohol effects that were often classified as biomedical, treatment, and prevention. The following five content areas and their criteria are:

1. *Biomedical or Medical* included: Alcohol and the central nervous system, Alcohol-related illnesses, Theories of causes (physiological), Heredity and congenital, Health consequences, Biomedical consequences of alcohol use/abuse, Fetal alcohol syndrome (physiological issues), Interaction of alcohol and other drugs, Genetics, Psychobiological, Medical consequences, Neuroscience, Actions on brain, Biochemical, Brain, Neurobiological, Neurons, Brain neurons, and Neurotoxicity.
2. *Epidemiology/risk factors* included: Patterns of use, Accidents and violence, Alcohol and highway safety, Special population groups, Social implications of alcohol abuse, Alcohol and pregnancy outcome (drinking patterns), and, Adverse social consequences.
3. *Treatment/Psychological Issues* included: Treatment of alcoholism and problem drinking, Psychological effects of alcohol, Family factors relating to alcoholism, Psychological/social causes of alcoholism, Diagnosis and treatment, Perception, Emotion, Sexuality, Barri-

       ers to treatment, Rehabilitation, Psychiatric co-morbidity, Psychological, and Social and developmental factors.

4.   *Prevention* included: Prevention of alcohol-related problems, Occupational alcoholism programming, Fetal alcohol syndrome (implications for prevention), Prevention and intervention, Screening/brief intervention, Reducing alcohol impaired driving, Community-based prevention, Alcohol advertising, and, Effects of changes in alcohol prices and taxes.

5.   *Health Services Research* included: Financing alcohol treatment services, Fiscal and human resources, Cost analysis of alcohol treatment, Alcohol treatment and health care costs, Alcohol and work productivity, Alcoholism and health insurance, Fiscal and human resources, and Economic costs of alcohol abuse.

As pointed out by Midanik (2004), lists of references at the end of each chapter were not included in the analysis. Further, some of the content in the ten reports did not fit into these five topic areas, for example, history and legal status of intoxication and alcoholism found in Volume 1, and were thus not included.

The percent coverage of each of the ten reports for the five content areas is shown in figure 5.1. What is striking about this figure is that each of the trend lines is very different with very unique trajectories. *Biomedical and Medical* aspects of alcohol abuse and alcoholism had considerably less coverage in the earlier reports but has clearly now become the dominant area. With two exceptions, the first four *Alcohol & Health* reports provided coverage of biomedical concerns proportionate to the other areas (with the exception of no coverage of health services research and low coverage of prevention in 1971; and high epidemiology coverage in 1971 and 1974). In 1983, these proportions changed dramatically. In the fifth report, attention to biomedical factors rose to 42 percent and rose to 48 percent by 1987. While coverage of this area decreased in 1990 and 1993, it has since increased in the last two reports to 46 percent in 1997 and 49 percent in 2000 representing more than twice the coverage of any other area.

Epidemiology reached its highest coverage in earlier reports with 43 percent coverage in 1974 but has decreased considerably since that time attention to the distribution of alcohol use and alcohol-related problems has steadily decreased so that only 14 percent of the last report published in 2000 was related to this area.

Coverage of the *Treatment/Psychological* area within these reports is variable. In the first report published in 1971, attention to alcohol

Figure 5.1
Coverage of the Five Content Areas, Alcohol and Health, 1971-2000

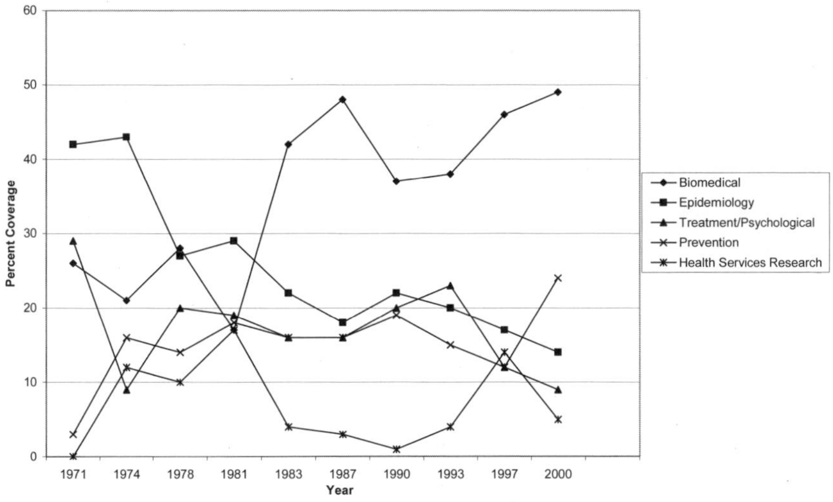

Source: Midanik 2004

treatment was included in 29 percent of the report. Since then, it has decreased in its coverage in the reports with only a 9 percent coverage rate in 2000—the same rate as in 1974.

Prevention as a content area was barely mentioned in the first *Alcohol & Health* report published in 1971 (3 percent coverage rate). Since that time, its coverage within the reports has risen to 24 percent in the most recent report that focused on a wide range of issues related to prevention including alcohol prices, taxes and advertising.

Health Services Research had no coverage in the first repot but has overall had the least coverage. Its peak was reached in 1981 with approximately 17 percent of the fourth report devoted to health services research issues. However, this has decreased considerably to approximately 5 percent slightly lower than *Treatment/ psychological*.

While the disproportionate attention to biomedical and medical issues related to alcohol abuse and alcoholism particularly in the last five reports precedes the movement of NIAAA to NIH, it has clearly been enhanced by the shift in institutional ideology. This attention placed on biomedical/medical (149 pages) is in sharp contrast to twenty-seven pages focused on treatment and psychological issues and the fourteen pages given for health services research. While prevention issues did get a larger focus in this latest volume (24 percent), it is clear that NIAAA has been expanding and showcasing its biomedical interests since 1983 and particularly in the 1990s. The ultimate result of this approach has been both an under-representation and a minimization of the importance of the other four areas separately and as important integrative factors.

Placement of Content in Reports

Beyond the coverage of specific topics in the ten reports to Congress and the disproportionate focus on biomedical areas for the last seventeen years, it is also important to note the order of these areas within each of the reports. Clearly areas that are deemed important and enhance the visibility of NIAAA will no doubt be placed early on in the reports; areas of less significance to the goals and interests of NIAAA would probably be discussed at the end of the reports.

The first or second chapters of almost all of the ten reports usually focuses on the distribution of consumption and alcohol-related problems in the U.S. population. Beginning with the first volume published in 1971 which contained only eight chapters, epidemiology is covered in the second chapter is entitled "Extent and Patterns of Use and Abuse of Alcohol." Biomedical issues ("Alcohol and the Central Nervous System" and "Alcohol-Related Illnesses") are included in chapters 3 and 4. Treatment is the focus of Chapter 6 and at least in this first volume, issues around prevention are not mentioned. The pattern of Epidemiology first, followed by biomedical, then treatment is maintained in the second volume but this time prevention occurs in a specified way under the heading "Problem Drinkers on the Job"—chapter 8. This clearly reflected NIAAA's interest in occupational alcoholism programs and early detection of problem drinkers at the workplace. Prevention as a specific identified area of interest appears first in the third report of *Alcohol & Health*, published in 1978. Once again, the ordering of *epidemiology, biomedical* and *treatment* remains the same with Prevention as a separate chapter (chapter 12) appearing last in the volume. It is also important to note that in this third volume, genetics emerged as a separate chapter (chapter 7) albeit with only three pages. While the ordering seems to make sense in these first three reports given that knowing the scope of the issue as described in the epidemiology chapter is important before discussing treatment and prevention, it is less clear why biomedical always follows epidemiology with a considerable number of pages devoted to it.

The fourth volume of *Alcohol & Health* published in 1981 went back to eight chapters with the somewhat standard order of content: epidemiology, biomedical, prevention and treatment. While the ordering remained the same for the fifth volume, in 1983 (with the one exception of treatment preceding prevention), genetics was now an expanded chapter that now directly followed epidemiology. This new focus on genetics and its import as demonstrated by where it is placed in the reports continued in the remaining volumes with the notable exception of the last report published in 2000. In this tenth report, genetics is replaced in its order by a substantial chapter on alcohol and the brain and its focus on neurobiology and neurobehavioral mechanisms. However, as this interest in genetics and neurobiological mechanisms has increased, prevention and treatment concerns are uniformly placed at the end of the reports designating a less important mission to NIAAA. Even in the latest issue of *Alcohol & Health* in which prevention is given an unprecedented coverage of 24 percent, it is still delegated to chapter 7 (of eight chapters) with treatment issues relegated to chapter 8.

This brief analysis of content placement within each of the ten reports of *Alcohol & Health* illustrates the importance given to some areas, for example, genetics, with much less attention or importance placed on other areas such as prevention or treatment—chapters which are placed at the end of the report.

Research Funding

The task of finding out how much money has been awarded over time by NIAAA for different categories of research projects (grants and contracts) might at first appear to be fairly easy. NIAAA is a federal agency under NIH, and, it awards public funds for research. Therefore, it might be assumed that this information is public domain and thus is readily available for anyone to see. However, this does not appear to be true. Even in the 1991 IOM report, there is a discussion of how difficult it was to obtain this information for their evaluation. In an effort to conduct a thorough examination of ADAMHA and its possible move to NIH by using specific case study areas, authors of the IOM report wanted to include a list of grants within each case study area and the amounts of the awards. The authors noted that there were two problems. The first was the format in which the data were available. A second problem was lack of information about the dollar amounts of some awards. For example,

> CRISP *[Computer Retrieval of Information on Scientific Projects]* contains no information on dollar amounts for intramural research projects. In addition, the CRISP listing of awards for subprojects of a multiproject grant (e.g., a clinical research center or a program project grant) shows the *average* of all subproject awards on the grant–not the actual amount for a particular subproject. This problem weighs heavily in any analysis, since centers and program project grants constitute a sizable portion of research funds. Furthermore, the costs of different kinds of research vary greatly, making comparisons difficult....(IOM, 1991, p. 49)

Thus, it is not an easy task to obtain the dollar amounts of research grants by category of grant even though public funds are being used.

For NIAAA this was not always the situation. The U.S. Department of Health and Human Services published at least two documents that provided some of this information on a yearly basis. One document entitled *Alcohol, Drug Abuse, Mental Health Research Grant Awards* appears to have been published until 1988. Within these documents several tables are presented that provide the total number of grants awarded by ADAMHA from 1950, the number of grants awarded separately by each agency for the current fiscal year, the amount of dollars awarded for all grants by state, the top ten institutions by dollars awarded for research, research scientist, fellowship and research training grants, and, a listing by state of each Investigator, the Project Title, the grant number, the program class code (PCC), and the amount of each award. What is key in this analysis is not only the award amount but the PCC. In the 1988 edition, these codes designated three separate divisions of NIAAA and six areas (five branches) within these divisions: *Division of Basic Research*: AL Alcohol Research Centers Branch, AM Biomedical Research Branch, AN Neurosciences and Behavioral Research; *Division of Biometry and Epidemiology*: AE Division of Biometry and Epidemiology; *Division of Clinical and Prevention Research*: AC Treatment Research Branch, AP Prevention Research Branch. With this information it is possible (yet tedious) to determine what proportion of funds are designated for Basic Research as compared to projects in the areas of biometry, epidemiology, clinical and prevention research. Clearly the PCCs have changed as new divisions and branches have been created or merged. However, it is interesting to note that in the 1977 edition, there is only one PCC for NIAAA named "Division of Research."

HHS also published another book that provided grant information by fiscal year entitled the *ADAMHA Data Book*. These smaller
reports also included chapters on AMAMHA's organization and history, the magnitude of the public health problems of alcoholism and
alcohol abuse, mental illness and drug abuse, ADAMHA's obliga-
tions by major program purpose (community programs, research,
training and program support) as well as ADAMHA's research obli-
gations and awards. Within this section, many charts and tables were
provided that allowed comparisons across agencies and within specific programs. However, in only one of these publications (FY 1983)
could any information be found that directly relates to the amount
of extramural research awards by program area. Thus, in 1983, 9.3
million dollars were devoted biomedical and genetic factors (36 percent), 6.5 million to alcohol-related problems and medical disorders (25 percent), 4.0 million for treatment (16 percent), 2.4 million
to epidemiology (9 percent), 1.8 million to prevention (7 percent) and
1.7 million to psychological and environmental factors (7 percent). Theo-
retically, it would be possible to look at trends in research awards at
NIAAA at least until 1992 if the information were readily available
from these ADAMHA publications. Unfortunately, the information
is difficult to locate and inconsistent in how and what is reported.

To track data on extramural research awards within categories of
grants for NIAAA following their movement to NIH in 1992, information was sought from the office of extramural research at NIH
(Wendy Baldwin, Deputy Director for Extramural Research, NIH,
and Robert Moore, director, Division of Statistics and Analysis, Office of Reports and Analysis, Office of Extramural Research, NIH).
Data were available in electronic format for all grants from 1990 to
2003 awarded by NIAAA; unfortunately, data on contracts were not
available. Program classification codes (PCC) were provided that
indicated which program branch was responsible for each grant.
Hence, it was possible to obtain data that could differentiate biomedical and medical awards from those focusing on epidemiology,
treatment, prevention, and health services research. The biomedi-
cal/medical area combined the Biomedical Research Branch and the
Neuroscience and Behavioral Research Branch. The other five areas were the Division of Biometry and Epidemiology, the Treatment
Research Branch, the Prevention Research Branch, the Health Services Research Branch, and Centers Program and Special Programs.

The Homeless Program and the Research Training Program Classification Codes have not been used by NIAAA since the early 1990s
and were therefore eliminated from this analysis. To adjust for price
changes between 1990 and 2003, the Gross Domestic Price (GDP)
Deflator (also known as the Implicit Price Deflator or the Implicit Price
Index.) was used. It incorporates price changes in all goods and services transactions in the United States, including the consumer, pro-
ducer, investment, government and international sectors. The GDP
Deflator was calculated using 2003 as the index year (U.S. Dept. of
Commerce, Bureau of Economic Analysis http://www.bea.doc.gov/).

This analysis extends Midanik (2004) by including the most recent data from 2003. Figure 5.2 presents the NIAAA Grant Awards
from 1990 to 2003 by the six program branches or divisions. Beginning in 1990, grant funding awarded for Biomedical and Neuroscience research has always been considerably higher than the other
areas, initially at approximately 49 million in 1990, to almost 147
million in 2003. An increase in biomedical funding began in 1996
with quite dramatic upturns in recent years. While there has been
increased funding in other areas such as Prevention (from 16 million in 1990 to approximately 39 million in 2003), there has been
substantial decreases in grant funding of treatment research (from
approximately 14 million in 1990 to approximately 9 million in
2003). With the exception of treatment, it appears that all the other

Figure 5.2
NIAAA Research Grant Awards by Branch/Division, 1990-2003

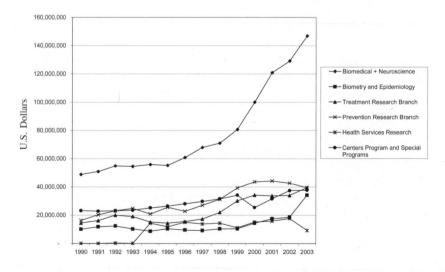

Figure 5.3
NIAAA Research Grant Awards by Branch/Division, 1990-2003
Number of Grants

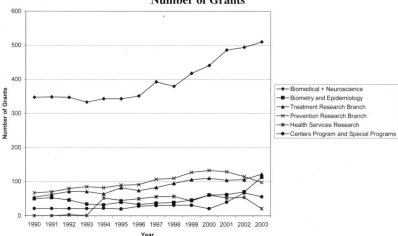

non-biomedical areas have merged at around 34 to 39 million in 2003. No other branch or division parallels the sharp rise in funding and the total amount of funding for Biomedical and Neuroscience.

These findings are parallel the number of grant awards given by NIAAA by branch or division during the same time period (figure 5.3). In 1990, 347 Biomedical/Neuroscience grants were awarded as compared to sixty-seven for Prevention, fifty-four for Treatment, fifty for Epidemiology, and twenty-one for the Centers and Special Programs. By 2003, Biomedical/Neuroscience increased to 510 funded grants, while the increases in the number of grants in other areas were more modest. In 2003, ninety-eight Prevention grants were awarded along with 122 Treatment grants, fifty-six Centers Programs and Special Program grants, 113 Epidemiology grants, and twenty-one Health Services Research grants. It should be noted that although Centers and Special Program grants are located in one category, the majority of the budget is allocated to Center grants. Historically, most of the Centers are biomedically focused. Thus, figures 5.2 and 5.3 underestimate the extent to which funding is allocated to biomedical research.

Conclusions

This chapter assessed the process of biomedicalization in the alcohol field in the U.S. as evidenced by NIAAA's mission, its Strategic Plan, the contents and their placement in NIAAA's ten

Alcohol & Health Reports to Congress, and NIAAA's funding patterns over the last 14 years. Taken together, there is strong evidence to suggest that biomedical factors dominant the research agenda of NIAAA as well as it mission. The consequences of this dominance are problematic.

Tesh (Tesh, 1988) strongly argues that values, beliefs and ideologies govern the work of both scientists and policymakers, and those who value "...individualism and positivism will probably make reductionistic analyses and come up with individualistic policies..." (p. 177). While Tesh (1988) was writing about disease prevention and not alcohol research per se, her conclusions are relevant to what has just been discussed. As biomedicalization, inherently focused on factors and conditions within the individual, prevails as a dominant lens through which alcohol problems are framed, policymakers and researchers will move towards individualistic solutions to social problems. Thus, environmental strategies such as access and availability, increased taxation, and restrictions on advertising will be minimized at best, and excluded at worst. Moreover, the research focus continue to be on the extreme cases of alcohol abuse, for example, alcoholism or alcohol dependence, and less on the larger, more pervasive alcohol issues that affect society such as underage drinking and drinking driving. If alcohol problems are increasingly seen as issues of alcohol dependent individuals, more attention will be placed on their identification and treatment and less on community and societal strategies that affect whole populations, thus including non-alcohol dependent individuals who may experience alcohol problem despite increasing evidence that these approaches can be an effective means to control alcohol use and alcohol-related problems (Babor et al., 2003; Edwards et al., 1994).

NIAAA's move to NIH symbolized and crystallized NIAAA (and the alcohol field) as a biomedical agency competing for funding with other agencies under NIH. Social science research has been relegated to "behavioral research," which appears to be a category defined by any research that is *not* biomedical. As such, this is a definition of social science research based on what it is not as opposed to what it actually is. Under this definition traditional treatment evaluation studies, for example, would be defined as behavioral research; yet, other areas are more questionable. Would, for example, population studies that examine the effects of licensing of outlets and its affects on alcohol use among underage drinkers be considered behavioral research? Would assessing the alcohol beverage industry's targeting of minority youth in their advertising also fall into that category as well? In other words, how broad is behavioral research?

One explanation for the lowered support of treatment and prevention studies is that these areas are funded in part by the Center for Substance Abuse Treatment (CSAT) and the Center for Substance Abuse Prevention (CSAP) both of which are located in the Substance Abuse Mental Health Services Administration (SAMHSA). Yet, both CSAT and CSAP maintain that they do not fund research. Thus while each center funds the development and evaluation of treatment and prevention programs, evaluation is not defined as "research," and thus it is impossible to estimate its funding levels.

Midanik (2004) discusses two additional indications of the biomedicalization process in the alcohol field. The first is the discontinuance by NIAAA of the ETOH Alcohol and Alcohol Problems Science Database that was widely used in the alcohol field particularly by social scientists. The second is the recent reorganization of NIAAA by its new director which combines Epidemiology and Prevention Research into one branch, eliminates Health Services Research as a separate branch or division, and has two branches (of the five) that are clearly biomedical (Neuroscience and Behavior and Metabolism and Health Effects). Both of these changes further signify a codification of biomedical research as the most important goal of the agency. It can be argued that the goals of any organization are determined by the "...interplay of power, the goals of the power wielders, and the interactions between the demand and supply of organizational services" (Hasenfeld, 1983, p. 95). For NIAAA, it is clear that the adoption of biomedical goals almost to the exclusion of other goals is compatible with the larger environment of NIH.

Biomedicalization is a very large phenomenon affecting the health field; alcohol studies and particularly alcohol research can be seen as a case study of this larger movement. Chapter 6 explores this issue from a cross-cultural perspective and examines if biomedicalization in the alcohol field is occurring in the same way in Sweden as in the U.S.

Appendix A
Documentation of Content Analysis
Alcohol and Health

*Volume 1,*1971: 99 pages, 3 blank pages, 15 ref pages = 81 pages

Biomedical: Alcohol and Central Nervous System = pp. 37-39 =
3 pages
Alcohol-Related Illnesses = pp. 45-54 = 10 pages
Theories about the Causes of Alcoholism: Physiologi-
cal = pp. 61-63 = 3 pages
Page 99= 1 page
TOTAL=17pages

Epidemiology:
Patterns of Use = pp. 21-35 = 15 pages
Proportions of Drinkers/Abstainers; Alcohol Con-
sumption, International Comparisons = pp. 9-18 =
10 pages
Accidents & Violence = pp. 41-42 = 2 pages
TOTAL=27pages

Treatment/Psychological:
pps: 39-40= 2 pages
pp 71-82 = 12 pages
Psychological/Social Causes of Alcoholism = pp. 64-
67 = 4 pages
Page 100= 1 page
TOTAL=19pages

Prevention.:   pps. 101-102= 2 pages
TOTAL=2pages

Health Services
Research: 0 Pages
TOTAL=0pages

TOTAL 65 PAGES CODED.
OMITTED**: History = pp. 5-8 = 4 pages
Legal status of Intoxication and Alcoholism = pp.
85-97 = 13 pages

Volume 2,1974: - 4 missing pages

Biomedical: Alcoholism: Heredity and Congenital Effects = pp.
61-66 = 6 pages
Some Health Consequences of Alcohol Use:
      Cancer = pp. 69-86 = 18 pages
      Heart = pp. 91-96 = 6 pages
      Liver disorders = pp. 98-101 = 4 pages
      Central nervous system = pp. 124-5 = 2 pages
TOTAL=36 pages

Epidemiology:
Alcohol Use and Misuse by Adults and Youth = pp.
1-32 = 32 pages
Alcohol and older Persons = pp. 37-46 = 10 pages
Alcohol and Highway Safety = pp. 127-41 = 15 pages
Alcohol and Mortality = pp. 104-21 = 18 pages
TOTAL=75 pages

Treatment/
Psychological: Trends in Treatment of Alcoholism = pp. 145-159= 15 pages
TOTAL=15 pages

Prevention: Problem Drinkers on the Job = pp. 169-180 = 12 pages
Licensure and monopoly states, licensing, state controlled provisions, definitions of alcoholic and
intoxicating beverages, prices, location, hours, and
eligibility, restrictions on persons, retail license limitations consumption and the lognormal curve, a coor-
dinated approach to regulation, standards of responsible drinking behavior = pp. 202-12 = 11 pages
Pages 197-201=5 pages
TOTAL=28 pages

Health Services
Research: Alcoholism and Health insurance = pp. 183-94 = 12 pages
Economic Costs of Alcohol-Related Problems = pp.
49-57 = 9 pages
TOTAL=21 pages

TOTAL 175 PAGES CODED.

Volume 3, 1978: 98 pages, 6 blank pages = 92 pages [no references in text of report]

Biomedical: Biomedical Consequences of Alcohol Use and Abuse
                 = pp. 25-37 = 13 pages
                 Fetal Alcohol Syndrome [p 44, double-coded in pre-
                 vention] = pp. 39-45 = 7 pages
                 Interaction of Alcohol and Other Drugs = pp. 47-52
                 = 6 pages
                 Genetic and Family Factors Relating to Alcoholism
                 = p 57 = 1 page
                 TOTAL=27 pages

Epidemiology:
                 Alcohol Use and Alcohol-Related Problems = pp. 1-
                 11, 15 = 12 pages
                 Special Population Groups = pp. 17-24 = 8 pages
                 Accidents, Crime, and Violence = pp. 61-66 = 6 pages
                 TOTAL=26 pages

Treatment/
Psychological: Pages 21-23= 3 pages
                 Treatment of Alcoholism and Problem Drinking = pp.
                 67-76 = 10 pages
                 Genetic and Family Factors Relating to Alcoholism
                 = pp. 58-60 = 3 pages
                 Psychological Effects of Alcohol = pp. 53-55 = 3 pages
                 TOTAL=19 pages

Prevention: Occupational Alcoholism Programming = pp. 77-83
                 = 7 pages
                 The Prevention of Alcohol Problems = pp. 93-98 = 6
                 pages
                 Fetal Alcohol Syndrome - Implications for Preven-
                 tion = p 44 = 1 page
                 TOTAL=14 pages

Health Services
Research: Financing Alcohol Treatment Services = pp. 85-91 =
                 7 pages

Economic Cost of Alcohol Related Problems = pp. 12-14 = 3 pages
TOTAL=10 pages

TOTAL 97 PAGES CODED.
Pages 21, 22, 23 coded in both Epidemiology and Treatment/Psychological.
Page 44 coded in both Biomedical and Prevention.

Volume 4, 1981: "Highlights" omitted from analysis [-13 pages], 178 pages

Biomedical: Biomedical Consequences of Alcohol Use = pp. 43-65
                TOTAL= 23 pages

Epidemology:
                Patterns of Alcohol Consumption = pp. 15-39 = 23 pages
                Social Implications of Alcohol Abuse = 81-93 = 13 pages
                TOTAL=36 pages

Treatment/
Psychological: Treatment and Rehabilitation = pp. 137-61
                TOTAL= 25 pages

Prevention: Prevention of Alcohol-Related Problems = pp. 103-16 = 14 pages
                Intervention = pp. 123-32 = 10 pages
                TOTAL=24 pages

Health Services
Research: Fiscal and Human Resources = pp. 169-89 = 21 pages
                Economic costs = p 93 = 1 page
                TOTAL=22 pages

TOTAL 130 PAGES CODED.
Page 93 coded in Epidemiology and Health Services.

Volume 5, 1983: 117 pages, omit "highlights"

Biomedical: Genetics = pp. 15-22 = 8 pages
Psychobiology = pp. 25-39 = 15 pages
Medical Consequences = pp. 45-59 = 15 pages
Effects of Alcohol on Pregnancy Outcome = pp. 69-75, 77 = 8 pages
TOTAL=46 pages

Epidemiology:
Epidemiology of Alcohol Abuse and Alcoholism = pp. 1-12 = 12 pages
Alcohol and Pregnancy Outcome - Drinking Patterns = pp. 75-76 = 2 pages
Adverse Social Consequences of Alcohol Use and Alcoholism = pp. 83-93 = 11   pages
TOTAL=25 pages

Treatment/
Psychological: Treatment: Emerging Trends in Research and Practice = pp. 100-16
TOTAL= 17 pages

Prevention: Prevention: A Broad Perspective = pp. 122-37 = 16 pages
Alcohol and Pregnancy Outcomes - Prevention = p 78 = 1 page
TOTAL=17 pages

Health Services
Research: Economic Costs of Alcohol Abuse = pp. 93-96
TOTAL=4 pages

TOTAL 109 PAGES CODED..
Page 75 coded in both Biomedical and Epidemiology.
Page 93 coded in both Biomedical and Health Services.

Volume 6, 1987: 142 pages

Biomedical: Genetics and Alcoholism = pp. 28-39 = 12 pages
               Psychobiological Effects = pp. 44-55 = 12 pages
               Medical Consequences = pp. 60-72 = 13 pages
               Fetal Alcohol Syndrome = pp. 80-94 = 15 pages
               TOTAL=52

Epidemiology:
               Epidemiology = pp. 1-20 = 20 pages
               TOTAL= 20 pages

Treatment/
Psychological: Treatment = pp. 120-136
               TOTAL=17 pages

Prevention: Prevention and Intervention = pp. 97-113
               TOTAL= 17 pages

Health Services
Research: Costs of Alcohol Abuse = pp. 21-23
               TOTAL=3 pages

TOTAL 109 CODED PAGES.

Volume 7, 1990: 280 pages minus 6 blank, minus 12 "introduction,"
total of 262 pages

Biomedical: Genetics/Environment = pp. 43-61 = 19 pages
               Neuroscience = pp. 69-94 = 26 pages
               Medical Consequences = pp. 107-28 = 22 pages
               Fetal Alcohol = pp. 144-53 = 10 pages
               TOTAL=77 pages

Epidemiology: pages 1-2=2 pages
               Epidemiology = pp. 13-37 = 25 pages
               Adverse Social Consequences = 163-73, 176= 12
               pages
               Fetal Alcohol Syndrome, Identifying High Risk Factors
               for FAS = pp. 139-44 = 4 pages
               TOTAL=43 pages

Treatment/
Psychological:

> Pages 3, 4, 6, and 7=4 pages
> Diagnosis and Assessment of Alcohol Use Disorders
> = pp. 181-99 = 19 pages
> Treatment = pp. 261-75 = 15 pages
> Fetal Alcohol Syndrome, Introduction, Clinical
> Studies: Followup Studies of FAS children, Effects
> of Lower Levels of Alcohol Drinking during Preg-
> nancy = pp. 139-43= 5 pages
> *TOTAL=43 pages*

Prevention: Page 4= 1 page
> Prevention = pp. 209-33 = 25 pages
> Early and Minimal Intervention = pp. 243-55 = 13
> pages
> FAS – Public Awareness and Policy = p 154 = 1 page
> *TOTAL=40 pages*

Health Services
Research: Economic Costs of Alcohol Abuse = pp. 174-75
> *TOTAL=2 pages*

TOTAL 205 CODED PAGES.
Page 5 omitted.
Pages 20 and 22 coded in both Epidemiology and Biomedical.
Page 134 coded as Biomedical and Prevention.

Volume 8, 1993: 349 pages minus 8 blank = 341 pages

Biomedical: Genetics = pp. 61-77 = 17 pages
> Actions on Brain = pp. 85-104 = 20 pages
> Neurobehavioral = pp. 113-22 = 10 pages
> Biochemical = pp. 147-58 = 12 pages
> Health and Body Systems = pp. 165-86 = 22 pages
> Fetal Alcohol Syndrome: Growth/Physical Deficits=
> pp. 207-10 = 4 pages;
> Characteristics/Benefits of Animal Models,
> Neuroanatomical Effects= pp 212-22=11 pages
> *TOTAL=96 pages*

Epidemiology:

        Epidemiology = pp. 1-31 = 31 pages
        Effects of Alcohol on Behavior and Safety = pp. 233-48 = 16 pages
        Epidemiology/Risk Factors for FAS=pp 203-6=4 pages
        TOTAL=51 pages

Treatment/
Psychological: Diagnosis and Treatment = pp. 319-42 = 24 pages
        Psychiatric Co-Morbidity = pp. 37-54 = 18 pages
        Psychological, Social and Developmental Factors = pp. 129-41 = 14 pages
        Pages 205-6= 2 pages
        TOTAL=58 pages

Prevention: Prevention = pp. 267-88 = 22 pages
        Screening/brief Intervention = pp. 297-311 = 15 pages
        FAS Prevention=p 211=1 page
        TOTAL=38

Health Services
Research: Economic issues = pp. 253-62
        TOTAL=10

TOTAL 253 CODED PAGES.
Page 204 coded in both Biomedical and Epidemiology.
Pages 205 and 206 coded in both Treatment and Epidemiology.

Volume 9, 1997: 398 pages minus 5 missing, 393 pages

Biomedical: Genetic Influences = pp. 33-48 = 16 pages
        Actions on Brain = pp. 65-90 = 26 pages
        Neurobehavioral = pp. 99-121 = 23 pages
        Health and Body Systems = pp. 131-69 = 39 pages
        Fetal Alcohol Syndrome = pp. 193-210, 215-30 = 34 pages
        TOTAL=138 pages

Epidemiology:

  Epidemiology = pp. 1-28, 33 = 29 pages
  Behavior and Safety = pp. 247-69 = 23 pages
  TOTAL=52 pages

Treatment/
Psychological: Treatment = pp. 337-63 = 27 pages
  Page 33= 1 page
  Psychological/Sociocultural Influences=pp 48-56=9 pages
  TOTAL=37 pages

Prevention: Fetal Alcohol Syndrome = pp. 211-14 - 4 pages
  Prevention = pp. 301-27 = 27 pages
  TOTAL=31 pages

Health Services
Research: Health Services Research = pp. 373-93 = 21 pages
  Economic Aspects of Alcohol Use and Alcohol-
  Related Problems = pp. 275-96 = 22 pages
  TOTAL=43 pages

TOTAL 301 CODED PAGES.
Page 33 triple-coded: Epidemiology, Treatment, and Biomedical.

Volume 10, 2000: 458 pages minus intros for each chapter (14 pages)
444 total

Biomedical: Alcohol and the Brain
       Neurons = pp. 69-76 = 8 pages
       Brain neurons = pp. 78-85 = 8 pages
       Brain = pp. 89-101 = 13 pages
       Neurobiological and Neurobehavioral = pp.
       107-23 = 17 pages
       Neurotoxicity = pp. 134-41 = 8 pages
       Pages 147-54= 8 pages

     Genetics
       Animal Genetic Studies = pp. 160-65 = 6 pages
       Recent Progress = pp. 169-77 = 9 pages

Medical Consequences
     Liver = pp. 198-207 = 10 pages
     Immune System = pp. 214-26 = 13 pages
     Cardiovascular = pp. 240-48 = 9 pages
     Women = pp. 253-55 = 3 pages
     Skeletal System = pp. 258-66 - 9 pages
     Breast Cancer = pp. 273-78 - 6 pages
     Prenatal Exposure= pp. 285-93 = 11 pages
     Pages 300-10=11 pages
TOTAL=149 pages

Epidemiology:

     Health Risks and Benefits=pp 3-10, 13-17=13 pages
     Over the Life Course=pp 28-45=18 pages
     Violence=pp 54-63=10 pages
TOTAL=41 pages

Treatment/
Psychological: Pages 11-12= 2 pages
     Treatment of Alcohol Dependence with Psych Problems=pp 444-48=5 pages
     Treatment of Alcohol Dependence with Medications=pp 451-58=8 pages
     Prenatal Exposure=pp294-95=2 pages
     Psychosocial Factors=pp 181-90=10 pages
TOTAL=27 pages

Prevention:

     Reducing Alcohol Impaired Driving=pp 375-91=17 pages
     Community-Based Prevention=pp 397-407=11 pages
     Alcohol Advertising=pp 412-23=12 pages
     Brief Screening and Intervention=pp 429-39=11 pages
     Issues in FAS Prevention=pp 323-32=10 pages
     Effects of Changes in Alcohol Prices and Taxes=pp 341-51=11 pages
TOTAL=72 pages

Health Services
Research: Economic and Health Services Perspectives = pp.
                355-61= 7 pages
                Pages 364-70= 7 pages
                TOTAL=14 pages

TOTAL 303 CODED PAGES.
Did not code glossary, pgs. 236-39.

References

Babor, T. F., R. Caetano, S. Casswell, G. Edwards, N. A. Giesbrecht, K. Graham, et al. 2003. *Alcohol: No Ordinary Comodity. Research and Public Policy*. New York, N.Y.: Oxford University Press.

Edwards, G., P. Anderson, T. F. Babor, S. Casswell, R. Ferrence, N. Giesbrecht, et al. 1994. *Alcohol Policy and the Public Good*. Oxford, U.K.: Oxford University Press.

Gordis, E. 1995. "The National Institute on Alcohol Abuse and Alcoholism: Past Accomplishments and Future Goals." *Alcohol, Health, and Research World* 19(1).

Hasenfeld, Y. 1983. *Human Service Organizations*. Englewood Cliffs, N. J.: Prentice-Hall, Inc.

IOM. 1991. *Research and Service Programs in the PHS: Challenges in Organization*. Washington, D.C.: National Academy Press.

Midanik, L. T. 2004. Biomedicalization and Alcohol Studies: Implications for Policy. *Journal of Public Health Policy* 25: 211-28.

NIAAA. 2002. (January). *NIAAA's Purpose*, from *http://www.niaaa.nih.gov/about/purpose.htm*

——. 2000. (October). *NIAAA Strategic Plan for 2001-2005*, 2004, from www.niaaa.nih.gov

———. 1986. *Toward a National Plan to Combat Alcohol Abuse and Alcoholism. A Report to the United States Congress*: Health and Human Services.

NIDA. 2000. (September). *NIDA Strategic Plan*, from *http://www.drugabuse.gov/StrategicPlan/index00-05.html*

Tesh, S. N. 1988. *Hidden Arguments: Political Ideology and Disease Prevention Policy*. New Brunswick: Rutgers University Press.

6

Biomedicalization and Alcohol Research
in Sweden

"Nordic health ministers have agreed for the first time to align their alcohol policies in order to reduce consumption. Meeting in Copenhagen on Monday (18 October 2004), the ministers of Finland, Sweden, Norway, Iceland and Denmark also agreed to act as a block during international negotiations in the World Health Organisation (WHO) and in the EU on alcohol policies. Alcoholic products are not like any other product and should be treated under special rules due to the social- and health-related effects of their consumption, the ministers said in a statement after their meeting. The Nordic ministers suggested higher taxes on alcohol in general, including on wine, and want to keep restrictions on quantities allowed to be transported by people travelling across borders." (European Public Health Alliance, October 29, 2004, http://www.epha.org/a/1505)

Introduction

The first five chapters of this book have approached the process of biomedicalization from the perspective of the U.S. In the general area of health and mental health, scholars who have written about how biomedical models have come to dominate the ways in which a specific area or condition is defined, diagnosed, prevented and treated have been predominantly from the U.S. (e.g., Clarke et al., 2003; Estes & Binney, 1989). The question to be asked then is how widespread is biomedicalization? Is what we are seeing in the U.S. contained within our borders or is biomedicalization so pervasive that it extends to other countries with very different histories of health care and mental health care policies?

To begin to address this issue in the alcohol field, I had the unique opportunity to conduct a study in Sweden, supported by a Fulbright Fellowship, to assess trends in alcohol research funding. Sweden was an ideal country to study in that its history and policies around health and mental health care and subsequently around alcohol policy are very different from the U.S.

Parts of this chapter were presented at the 30th Annual Alcohol Epidemiology Symposium of the Kettil Bruun Society for Social and Epidemiological Research on Alcohol, Helsinki, Finland, June 2004.

Historically, Swedish alcohol policy has been strongly influenced by the temperance movement; yet, by a very narrow margin, a rationing system, known as the Bratt system, was created instead of formal prohibition in 1917. Under this system, ration books were issued to adult men (with smaller rations to unmarried women and younger males) that allowed up to four litres of distilled spirits per month to be purchased in one monopoly shop (Holder, 2000). When this system was discontinued in 1955, the alcohol retail monopoly, Systembolaget, was retained and continued its monopoly on off-premise retail sales. Domestic distilled spirits production and the importing of spirits, wine, and strong beer was handled by Vin & Sprits, the state-owned Central Wine and Spirits Corporation (Karlsson & Österberg, 2002).

Over time, alcohol control in Sweden has changed considerably most notably with Sweden's entrance into the European Union (EU) in 1995 (Holder et al., 1998). Under the new Alcohol Act, Sweden no longer has monopoly control over the production, import, export, and wholesale aspects of alcohol sales that were previously under the purview of Vin & Sprits. Systembolaget, however, maintained its monopoly over retail sales. Kühlhorn and Trolldal (2000) argue that with Sweden's entrance into the EU, two important principles of Swedish alcohol policy were affected: "the high-price policy and the so-called principle of elimination of private profit motives from the alcohol market" (p. 40). As alcohol controls have been loosening, Sweden is correspondingly experiencing an ongoing increase in recorded alcohol consumption (Leifman & Gustafsson, 2003). Recent data indicate a 74 percent increase in wine and a 70 percent increase in strong beer consumption from 1996 to 2003 based on survey data (Ramstedt & Gustafsson, 2004). Sweden's current alcohol policies continue to be affected by lower taxes on alcoholic beverages in neighboring EU states such as Finland and Denmark as well as dramatic increases in travellers' allowances among EU states that permit large amounts of lower priced alcohol to be brought into Sweden. Data from 2003 show that 22 percent of the alcohol purchased is from other countries (Ramstedt & Gustafsson, 2004). As the quote at the beginning of this chapter indicates, the Nordic countries are attempting to form a voting block within the EU to maintain their alcohol control policies even though they are not in line with other EU states. Given the many changes currently occurring in Sweden's alcohol policy, the ability to conduct robust research projects that examine these shifts and their effects is critical. Thus, the issue of funding alcohol research from a wide array of perspectives becomes increasingly important.

Biomedical research appears to have become steadily more important in Sweden in the 1990s. While the membership is cross-disciplinary in the 300 member Swedish Alcohol and Drug Research Society (SAD), biomedical research has become dominant in it, reflected by the fact that the present chair and a majority of the board are biomedical or clinical researchers. A government committee on the future of alcohol research remarked in 1995 that "while there is an active tradition of social alcohol research in Sweden, it has been overshadowed in the previous ten years by biological and clinical studies" (Alkoholforskingsutredningen, 1995). In response to the committee's report, the Centre for Social Research on Alcohol and Drugs (SoRAD) was established at Stockholm University in 1999.

Clearly there continues to be much concern in Sweden about alcohol use and its potential problems given the recent changes in policy. In 2001, a national action plan to prevent the harms, medical and social, of alcohol was adopted. The objective of this plan is to reduce the total consumption of alcohol and reduce harmful drinking patterns. The sub-goals of this plan as outlined on the website www.alkoholkommitten.se are: alcohol should not be consumed in traffic situations, at workplaces or during pregnancy, children should grow up in an alcohol-free environment, the age of onset of alcohol consumption should be delayed, intoxication should be reduced, there should be more alcohol-free environments, illegal dealing in alcohol should be eliminated. An allocation of SEK 680 million (approximately 52 million U.S. dollars) over four years has been devoted to fund this plan. Most of the funds, SEK 435 million (approximately 33 million U.S. dollars), will be given to the municipalities to reinforce actions. The remaining funds, SEK 245 million (approximately 19 million U.S. dollars), will be used to increase support for developing voluntary organizations and research.

The purpose of this chapter is to describe how alcohol research has been funded in Sweden from 1990-2003 in terms of its disciplinary emphases and compare it to what has been found in the U.S. Unlike the U.S., alcohol research in Sweden is funded by many agencies including multiple public as well as private sources. Thus, the task of determining the proportion of funding allocated to different types of alcohol research is more challenging. Data from the major alcohol funding agencies in Sweden were examined to assess the relative financial support that has been given to different types of alcohol research in Sweden over the last fourteen years. Second, data from in-depth interviews with representatives from agencies that fund alcohol research, other alcohol-related agencies, and biomedical and social science alcohol researchers were used to supplement the funding data to assess the direction of alcohol research funding in Sweden and its effects on treatment and prevention services.

Alcohol Research Funding

Based on interviews and discussions with individuals in the alcohol field, multiple sources of alcohol research funding were identified. Of these sources, data on alcohol research funding were obtained from eleven sources described briefly below.

1. *Vetenskapsrådet, The Swedish Research Council*: As stated on their website (http://www.vr.se) "The Swedish Research council bears national responsibility for developing the country's basic research towards attainment of a strong international position. The Council has three main tasks: research funding, science communication and research policy." As such, it is organized into four separate scientific councils: humanities and social sciences, medicine, natural and engineering sciences, and educational science. To obtain information on alcohol research funding, we sought data from the humanities and social sciences (Hum/SS) and medicine (Med) councils. (The natural and engineering sciences and the educational sciences councils have not funded alcohol research.) Data on alcohol research projects funded by the Hum/SS Council were obtained directly from a representative of the Council; data on alcohol research projects funded by the Med Council required looking through annual reports kept at Council. Based on title and abstract, we determined which projects were alcohol-related. Any project that examined alcohol as part of its research agenda was included.

2. *Forskningrådet för Arbetsliv och Socialvetenskap (FAS), Swedish Council for Working Life and Social Research*: FAS was established in 2001 when two other councils, the Swedish Council for Social Research and the Swedish Council for Work Life Research, were merged. Its mission as stated is: "To promote the accumulation of knowledge in matters relating to working life and the understanding of social conditions and processes through Promotion and support of basic and applied research. Identification of important research needs. Dialogue, dissemination of information and transfer of knowledge. Promotion of cooperation between researchers both nationally and internationally, particularly in EU programmes." (http://www.fas.forskning.se/en/).

3. *Systembolagets fond*: Systembolagets Råd För Alkoholforskning (SRA) is a state-owned company that also is also the alcohol retail monopoly in Sweden. SRA funds research projects with a focus on prevention of alcohol injuries. This Council is separate from the Systembolaget. It spends approximately 3 million SEK each year on research. The administration of the Council is handled by CAN (Swedish Council for Information on Alcohol and other Drugs). (http://www.can.se/SRA/SRA.asp)

4. *Riksbanken jubileumsfond*: This funding source was established in 1962 based on a donation from Sveriges Riksbank (Sweden's central bank). The foundation supports Swedish research has today funded approximately 4.5 billion SEK. It is the largest research financier for social science and humanities outside of funds available through universities and colleges. Its board includes members from universities, colleges, The Swedish Parliament and economic administration. To chart the need for different research areas and stimulate research and information exchange in different areas, which are seen as important but have not gotten sufficient attention, it sometimes creates special "area groups" and alcohol research is one of them.

5. *Ministry of Health & Social Affairs:* The areas of responsibility of the Ministry of Health and Social Affairs relate to social welfare: financial security, social services, medical and health care, health promotion and the rights of children and disabled people.

6. *Private Insurance Companies:* Three private insurance companies, AFA, Salusansvar and Trygg-Hansa provided some research funding for both social science and biomedical research.

7. *Vin & Sprits:* Vin & Sprits (also known as Vin- och sprithistoriska museets stipendium: The Historical Museum of Wines and Spirit's Grant) encourages scientific research on alcoholic beverages and is distributed yearly. It funds research within Swedish alcohol culture with a focus on traditions, attitudes and drinking patterns. Research can conducted on technique and manufacturing history as well as in trade and politic (www.vinosprithistoriska.se).

8. *Foundations:* Three foundations administered by the Swedish Society of Medicine (Svenska Läkaresällskapet, Söderström-Königska, Socialstyrelsen).

9. *Fokhälsoinstitutet (FHI), National Institute of Public Health*: The National Institute of Public Health (NIPH) in Sweden is a national centre of excellence for the development and dissemination of methods and strategies in the field of public health. It is responsible for comprehensive cross-sectoral follow-up and evaluation of national public-health policy. It exercises supervision in the areas of alcohol, drugs and tobacco, and its activities are conducted on the basis of scientific evidence (http://www.fhi.se/default 1417. aspx).

10. *Traffic Safety:* Three agencies concerned with traffic safety and drunk driving (Vägverket, Swedish Probation Service, Swedish Road Administration).

11.   *Statens Institutionsstyrelse (SIS):* "The National Board of Institutional Care (SiS) exists for those who are most disadvantaged. In forty-nine LVM homes and special approved homes in various parts of Sweden, we look after young people and adults who, in various ways, have "gone off the rails". Most young persons and substance abusers in SiS institutions have been placed there without their consent because they are in danger of injuring themselves or of ruining their lives, but there are also cases of voluntary admission under the Social Services Act" (http://www.stat-inst.se/zino.aspx?articleID=87).

Of these eleven sources, six funded both biomedical and social science research (Vetenskapsrådet, Systembolaget's fond, Riksbanken jubileumfond, Ministry of Health and Social Affairs, private insurance companies, and foundations), four funded social science research only (FHI, FAS, the drunk driving funding agencies, and SIS), and one funded biomedical research only (Vin & Sprits).

Other sources of alcohol research funding were identified but are not included in this analysis for several reasons. In some cases, data were not available to us (e.g., pharmaceutical companies) or were difficult to locate (research units in municipalities, funding administered through hospitals from county councils). We were also not able to determine specific amounts of funding allotted to Sweden based on funding sources outside of Sweden such as the EU Commission and the U.S. National Institutes of Health. One outside funding source, NOS-HS (Joint Committee for Nordic Research Councils for Humanities and the Social Sciences), is funding an alcohol research project beginning in 2004, and thus is also not included in our analysis. Finally, we did not include as "alcohol research" many treatment studies funded by CUS, Centrum för utvärdering av socialt arbete, National Board of Health & Welfare and SIS, Statens Institutionsstyrelse (only one study funded by SiS that is alcohol-focused is part of the trend data), because most of these studies focused on compulsory treatment issues generally among drug users. While clearly alcohol is also used by these populations, it is not a focus of this research, and thus we did not include these studies in our analyses.

For each source from which data were available, the amount funded and the number of grants funded by type of alcohol research were collected from 1990-2003. These dates were chosen because they provide a long enough timeframe to capture shifts in research funding, they cover five years prior to and nine years following Sweden's entrance into the EU in 1995, and the years match those used in the U.S. analysis (Midanik, 2004) and thus can be compared. In some instances, only total budgets were available to us for multi-year grants. In those situations, we divided the total budget by the number of years for each grant and used this average for each year of the grant.

Figures 6.1 and 6.2 present the total of Swedish alcohol research funding in SEK and by number of grants from 1990-2003. (Table 6.1 presents these data in more detail by source of funding.) Roughly, the lower limit on figure 6.1, 5 million SEK, would be equivalent to approximately 550,000 Euros or 660,000 U.S. dollars. The upper limit, 40 million SEK, would be approximately 4.4 million Euro or 5.3 million U.S. dollars. Figure 6.1 indicates a significant rise in alcohol research funding from 1998 to 2001. This represents an increase of over 300 percent during that time: approximately 12 million SEK to 39.5 SEK. Interestingly, the number of grants appears to be fairly stable during that time period with a drop in 2003.

Figures 6.3 and 6.4 present alcohol research funding in Sweden by SEK and by number of grants funded for biomedical/medical grants and for social science, epidemiology and psychology combined. Tables 6.1 and 6.2 present the raw data by source of funding and number of grants. Both types of research appear to have been funded at similar levels until the late 1990s. Beginning in 1998, there is a large increase in funding for social science, epidemiological and psychological research; by 2001, funding for biomedical research is at approximately 6 million as compared to 34 million for the combined group. Funding of both types of alcohol research has declined in the last 2 years. The number of biomedical alcohol research grants that were funded declined slightly during the early 1990s then began to increase. In 2002, 28 biomedical grants were awarded (the highest number during the study period). For social science, epidemiology, and psychology, the number of grants has always been higher than biomedical, reaching its peak in 1995 with 58 grants. Both types of research show a decrease in 2002.

Figure 6.1
Total Alcohol Research Funding in Sweden (SEK): 1990-2003

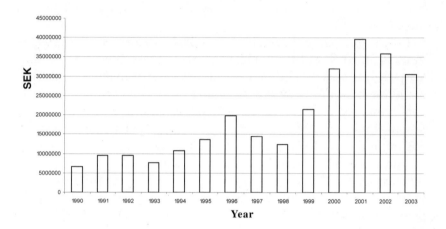

Figure 6.2
Total Number of Alcohol Research Grants Funded in Sweden: 1990-2003

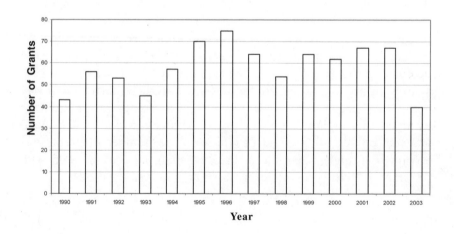

Figure 6.3
Alcohol Research Funding in Sweden (SEK) by Types of Grants: 1990-2003

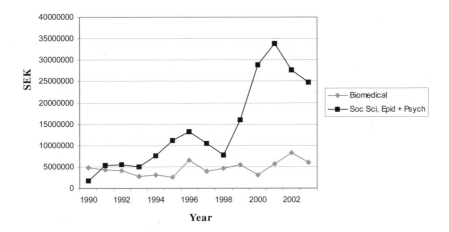

Figure 6.4
Alcohol Research Funding in Sweden (Number of Grants) by Types of Grants:
1990-2003

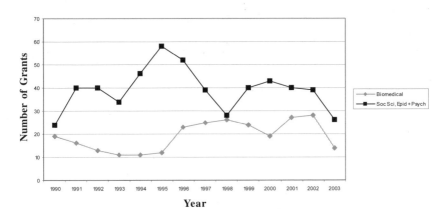

Table 6.1
Alchohol Research Funding in Sweden (SEK) by Funding Source: 1990-2003

Biomedical

	1990	1991	1992	1993	1994	1995	1996	1997	1998	1999	2000	2001	2002	2003
Vetenskapsrådet	2630000	2423000	2889000	2440000	2247000	1709000	4941000	2861000	3424000	3848000	2036000	3209000	3455000	2872000
Systembolagets fond	255500	255500	255500	255500	255500	255500	1030000	1105000	1200000	1200000	1100000	1275000	1300000	
Riksbanken jubileumfond	1842000	1665000	946000		600600	600600	600800							
Ministry of Health and Social Affairs												1000000	650000	1000000
Insurance Companies													2000000	1000000
Vin & Sprits										400000			850000	850000
Foundations												250000	40000	205000
Total	4727500	4343500	4090500	2695500	3103100	2565100	6571800	3966000	4624000	5448000	3136000	5734000	8295000	5927000

Social Science

	1990	1991	1992	1993	1994	1995	1996	1997	1998	1999	2000	2001	2002	2003
Vetenskapsrådet						582000	610000	617000					365000	570000
Systembolagets fond	1792222	1792222	1792222	1792222	1792222	1792222	1070000	895000	540000	800000	920000	725000	1700000	990000
Riksbanken jubileumfond														
National Institute of Public Health, FHI					2625533	6190509	6844733	5174900	2154277	4410000	3750000	3750000		1000000
Ministry of Health & Social Affairs					350000						10020000	15440000	9715000	8305000
Insurance Companies							95000	130000	50000	887500	787500			100000
FAS		3456000	3691000	3112400	2797000	2554713	4591938	3648860	4698650	7582415	11205909	12311450	15032800	13188500
Traffic Safety										2000000	1800000	1500000	690000	500000
Foundations											50000	100000	194000	54000
SiS-Statens Institutionsstyrelse									316666	316666	316666			
Total	1792222	5248222	5483222	4904622	7564755	11119444	13211671	10465760	7759593	15996581	28850075	33826450	27696800	24707500

Table 6.2
Alchohol Research Funding in Sweden (Number of Grants) by Funding Source: 1990-2003

Biomedical

	1990	1991	1992	1993	1994	1995	1996	1997	1998	1999	2000	2001	2002	2003
Vetenskapsrådet	11	9	8	8	7	8	7	10	10	11	8	12	13	8
Systembolagets fond	3	3	3	3	3	3	15	15	16	11	11	13	9	
Riksbanken jubileumfond	5	4	2		1	1	1						2	1
Ministry of Health and Social Affairs												1	1	1
Insurance Companies													2	1
Vin & Sprits														
Foundations										2		1	1	3
Total	**19**	**16**	**13**	**11**	**11**	**12**	**23**	**25**	**26**	**24**	**19**	**27**	**28**	**14**

Soc Sci, Epid + Psych

	1990	1991	1992	1993	1994	1995	1996	1997	1998	1999	2000	2001	2002	2003
Vetenskapsrådet						1	2						1	1
Systembolagets fond	24	24	24	24	24	24	14	12	8	9	10	6	16	
Riksbanken jubileumfond														1
National Institute of Public Health, FHI					13	26	25	17	7	12	9	9		1
Ministry of Health & Social Affairs											7	9		8
Insurance Companies					1		2	2	1	3	2		4	1
FAS		16	16	10	8	7	9	8	11	15	13	14	15	13
Traffic Safety											1	2	3	1
Foundations														
SiS-Statens Institutionsstyrelse									1	1	1			
Total	**24**	**40**	**40**	**34**	**46**	**58**	**52**	**39**	**28**	**40**	**43**	**40**	**39**	**26**

Key Informant Interviews

To supplement the trend data and to obtain more information about the funding process of alcohol research in Sweden, qualitative interviews were conducted with ten individuals. Half of the interviewees were alcohol researchers themselves and represented a wide range of disciplines including biomedical and clinical research. Three of these researchers worked in hospital research settings (Karolinska Institute) and two of the five researchers were directly involved in the funding of research. Additionally, four individuals who currently or formerly worked at funding agencies were interviewed (Ministry of Health and Social Affairs, Mobilization against Drugs, Vetenskapsrådet, and the former director of Systembolaget). Those who were alcohol researchers were asked primarily about the type of research they conduct, their funding sources, their general impressions about the funding process and if they perceived any changes beginning in the 1990s, their general feelings about the proportion of research funding available to them and to other types of research in the alcohol field, and whether they have seen shifts in alcohol research funding particularly since Sweden entered the EU. Those who represented funding agencies were asked specifically about the process of distributing grant funds, the review process, and funding trends over time.

There are several themes that arose in the interviews related to alcohol research and its funding that seem relevant. First, most of the researchers who were interviewed pointed out that they were not funded well enough by any one agency to do solid large-scale research. This is not surprising given the diffuse ways in which alcohol research is funded in Sweden. Further, interviews with individuals representing funding agencies indicated that increasingly these agencies are using a strategy often referred to, in the U.S., as "thinning the soup." That is, funding agencies prefer to use their money to fund more grants at significantly smaller levels than what was requested for any one project. Thus, it is unusual for any one researcher to obtain the total funds needed and requested from a specific funding agency. More likely funders will provide considerably less money requested in order to fund established researchers as well as to fund new researchers in the field and hopefully encourage them to develop their research ideas. As a result, more researchers receive funding but at smaller amounts.

A second area that is important to note is that no one agency monitors research money in the alcohol field. Because funding arenas are diffuse, it is very difficult to track which sources give money, to whom, how much, and over what time period. While several interviewees indicated that funders do talk to each other, there is no overall monitoring or coordination effort. Also, how funding is categorized varies by agency. There appears to be no consistent manner in which types of research, particularly non-medical or non-biomedical research, is handled. While Vetenskapsradet has a Humanities/Social Science Council, studies of social factors involving alcohol dependent persons or populations studies of alcohol consumption are funded under the Medical Council if it is submitted by someone in a medical research setting. This differs from other funding mechanisms such as Systembolagets fond that differentiates social science research, from psychology, from epidemiology, from medical research based on the content of the study.

There appears to be a considerable amount of tension among researchers concerning how research gets funded—either by a competitive mechanism or by sole source or contract arrangements. In Sweden, it appears that social science research is more often directly requested and funded by government agencies while biomedical research is more likely to be funded by competitive grant review. Because multiple state agencies fund Swedish alcohol research, there is less consistency across funding agencies on how grants are written and how they are reviewed. As was pointed out to me by those interviewees involved in the funding process, Sweden does not have an U.S. National Institutes of Health equivalent, and there is some thought that eventually Sweden may move in this direction.

Finally, most funded treatment research is focused on compulsory treatment issues in which individuals who have problems with illicit drugs are the focus. For example, from 1994-2002, Statens Institutionsstyrelse provided approximately 31.5 million SEK for research on youth in compulsory treatment and 25.5 million SEK for adult compulsory treatment. However, none of the projects focus on alcohol use. While other treatment research does occur that addresses alcohol-dependent clients, it is difficult to monitor funds in this area. The Addiction Center in Stockholm in partnership with Karolinska Institute conducts many treatment studies (alcohol, illicit drugs and combined) but the funding for these projects is distributed to different departments and is separately monitored making it difficult to track funds over time.

Discussion

These findings indicate that biomedicalization of the alcohol field, at least the extent to which it has been experienced in the U.S., is not occurring in Sweden. Further, it appears that the efforts made to compensate for an imbalance of more biomedical research in the early 1990s by creating a center that focuses on social research on alcohol and drugs (SoRAD) in 1999 were successful. Thus, as the data indicate, the proportion of funding for social science, epidemiological and psychological (non-biomedical) research has exceeded biomedical research increasingly from 1996 to the present.

There are several explanations for these findings. First, Sweden has a long history of social medicine. That is, the Swedish medical profession has historically focused on the social contexts of disease as well as its biomedical causes. Moreover, in the mid 1600s, Sweden's health care became separated from Poor Laws that stigmatized use of health services. By the mid 1800s, Sweden reorganized government to include nationalized health care "...pre-dated by eighty years England's nationalization of health and mental health services" (Armor, 1981, 63). As a result of this nationalized effort and continuing emphasis of the important role of the social world within medicine, researchers from different disciplines may be less competitive or have fewer contentious issues concerning the conflux of biomedical and social factors that are most important to study in the alcohol field. Based on the funding data, it is interesting to note, that in some cases, how a grant is classified is more often due more to the discipline of the researcher or where he or she is working than the specific content or type of project itself. Thus, epidemiological studies which may involve looking at social factors of alcohol dependent persons are funded under "medicine" in Vetenskapsradet, the Swedish Research Council, whereas they are not classified as medical when funded by Systembolagets fond. This suggests that medicine as conceptualized by Vetenskapsradet, and perhaps by Sweden in general, is more broadly defined.

Second, "purely" based biomedical research may be less favored in Sweden and left to other countries such as the U.S. with bigger research budgets to conduct. One clinical researcher noted that research budgets that are based on agreements between the county and the hospital clinics require that biomedical researchers work with these local clinics in their research projects in order to gain access to these funds. This suggests that biomedical researchers may increasingly blend their research with clinicians so that their findings can be readily applied.

Third, Sweden's entrance into the EU and national concerns with alcohol policy as an EU member may be a major impetus for more research being funded that focuses on alcohol policy and alcohol treatment. As the quote at the beginning of the chapter suggests, Sweden and the other Nordic countries are forming an alliance to counteract some of the "neutralization" of their alcohol control policies. Their work will rely heavily on research data that assess the effects of these restrictive policies given the existing "looser" policies of the EU.

There are several limitations to this research that need to be noted. First, some funding sources for alcohol research in Sweden are difficult if not impossible to obtain such as information on specific research projects funded through hospital clinics (in conjunction with the counties). Information on Foundations that fund alcohol research, most of which are administered by the Swedish Medical Council, were only available from 1999-2003. From the interviews with alcohol researchers, it became apparent that some alcohol research is done "through the backdoor." For example, many drug research products, ranging from pharmacological studies to treatment studies of illicit drug use, may also include alcohol, but alcohol is often not included in the title or the abstract. Thus, it is difficult to monitor these projects. Finally, we were unable to determine how many grants were submitted or rejected in the alcohol area for these agencies.

In conclusion, this research calls for further inquiry concerning the direction of alcohol research in Sweden. Clearly, Sweden is undergoing significant changes in its restrictive alcohol policy—a change that will no doubt have major consequences on the harms and problems associated with increases in alcohol consumption. How much encouragement should be placed on biomedical research versus social science, psychological and epidemiological research is unclear. In assessing priorities for research in alcohol and drugs in Canada, Room and Rehm (2003) suggest that biological researchers should consider focusing on gaps in knowledge that are directly applicable to social problems, for example, how to measure impairment from cannabis and driving as opposed to the perhaps more prestigious research agendas such as the etiology of addiction. Given limited and decreasing budgets and the inducement of county/hospital clinic funds for biomedical researchers who are directly affiliated with addiction clinics in Stockholm, this may be the direction of biomedical research in Sweden in the future. What is clear is that it will be important to continue to monitor the types and amounts of alcohol research grants that are funded in Sweden. This leaves just one important unanswered question: *Who will fund research on alcohol research funding?*

Acknowledgments

The author wishes to acknowledge Fabian Sjö for his assistance in obtaining and interpreting the funding data, Drs. Robin Room and Börje Olsson for their ideas and support of this project, and the staff of the Centre for Social Research on Alcohol and Drugs (SoRAD) of Stockholm University for their ongoing encouragement and help. Finally, I would like to thank the following people for participating in the interviews: (1) Jan Blomqvist, associate professor, research director, Research & Development, Social Services Administration, Stockholm; (2) Vera Segraeus, adjunct professor, SoRAD, former research director, National Board of Institutional Care (SiS); (3) Fred Nyberg, director of research, Mobilisering mot narkotica (MOB), The Swedish National Drug Policy Coordinator; (4) Lars Terenius, professor in clinical neuroscience, Karolinska Institute, Solna; (5) Bjorn Hibell, director, Swedish Council for Information on Alcohol and Other Drugs (CAN); (6) Marcus Heilig, professor, Department of Clinical Neuroscience, Karolinska Institute, Huddinge; (7) Gert Knutsson, Ministry of Social Affairs; (8) Bo Öhngren, Vetenskapsrådet; (9) Hans Bergman, professor, Department of Clinical Neuroscience, Clinical Alcohol and Drug Addiction, Karolinska Institute, Solna; and, (10) Gabriel Romanus, Swedish Parliament, former head of Systembolaget.

References

Alkoholforskningsutredningen (1995) Forskning om alcohol för individ och samhälle [Research on alcohol for the individual and society]. Stockholm, Folhälsoinstitutet 1995:49.

Armour, Philip K. (1981) The Cycles of Social Reform. Mental Health Policy Making in the United States, England, and Sweden. Washington, DC: University Press of America.

Clarke, A., Shim, J. K., Mamo, L., Fosket, J.R., and Fishman, J.R. "Biomedicalization: Technoscientific transformations of health, illness, and U.S. biomedicine," *American Sociological Review* 68 (2003): 161-194.

Estes, C.L. & Binney, E.A. (1989). The biomedicalization of aging: Dangers and dilemmas. *Gerontologist, 29,* 587-596.

Holder, Harold. D., Kühlhorn, Eckart, Nordlund, Sturla, Österberg, Esa, Romelsjö, anders, Ugland, Trygve 1998 *Euopean Integration and Nordic Alcohol Policies. Changes in alcohol controls and consequences in Finland, Norway and Sweden, 1980-1997.* Brookfield: Ashgate.

Holder, Harold D. History of Swedish national policies about alcohol: 1855-1995. In: Holder, H.D. (Ed) *Sweden and the European Union: Changes in National Alcohol Policy and their Consequences.* Stockholm: Almqvist & Wiksell, 2000, pp 15-27.

Karlsson, Thomas & Österberg, Esa Sweden. In: Österberg, Esa, Karlsson, Thomas (Eds) *Alcohol Policies in EU Member States and Norway. A Collection of Country Reports.* Stakes, Alcohol and Drug Research, Helsinki, Finland, 2002, pp 383-406.

Kühlhorn, Eckart & Trolldal, Björn The process of changes during Sweden's integration into the European Union. In: Holder, H.D. (Ed) *Sweden and the European Union: Changes in National Alcohol Policy and their Consequences.* Stockholm: Almqvist & Wiksell, 2000, pp 29-41.

Liefman, Håkan & Gustafsson, Nina-Katri (2003) *En skål för det nya millenniet. En studie av svenska folkets alkoholkonsumption i början av 2000-talet.* Forskningsrapport nr. 11, SoRAD, Stockholm.

Midanik, L.T. (2004) The biomedicalization of alcohol studies: Implications for policy. Journal of Public Health Policy, 25, 211-228.

Ramstedt, Mats & Gustafsson, Nina-Katri (2004) Alkoholkonsumtionen i Sverige 1996-2003. Paper presented at SoRAD Day Seminar, May, 2004.

Room, Robin & Rehm, Jurgen (2003) Research priorities for Canada in alcohol and drugs. Center for Addiction and Mental Health (CAMH), Toronto, Ontario, September.

Sulkunen, Pekka, Sutton, Caroline, Tigerstedt, Christoffer, Warpenius, Katariina (Eds) 2000 *Broken Spirits. Power and Ideas in Nordic Alcohol Control.* Nordic Council for Alcohol and Drug Research (NAD) Publication No. 39, Helsinki, Finland.

7

Biomedicalization and the Social Sciences[1]

Introduction

No doubt it is common as one approaches the end of a book to feel somewhat overwhelmed concerning what conclusions can be drawn or directions taken at this juncture. That biomedicalization is occurring in many fields is not necessarily either negative or undesired. Thus, one cannot conclude that steps should, or even can, be taken to stop this movement. Rather, the focus of this book is to ponder a larger question of how this trend can be interpreted and balanced with other research approaches so that there are ample opportunities to study social issues from multiple perspectives. One way to begin to undertake this process is to highlight the achievements brought to the alcohol field by social scientists within the context of biomedicalization. The next section of this chapter will explore this in more detail followed by a discussion on ways in which social scientists and biomedical researchers can work collaboratively on social problems.

Contributions of Social Science in an Era of Biomedicalization

It can be argued that one reason the recent focus in the alcohol field has been primarily biomedical is because that is precisely where "breakthroughs" occur as expressed in leading journals, newspaper articles, and other media. It is groundbreaking news when a new medication seems to succeed, even on small clinical samples, or when genes for any disease are "discovered," given the focus in recent years on the human genome project and its coverage in the press. Follow-up stories on medication treatments or genetic break-

Parts of this chapter were presented at the 29th Annual Alcohol Epidemiology Symposium of the Kettil Bruun Society for Social and Epidemiological Research on Alcohol, Krakow, Poland, June 2, 2003. It was first published in *Social Science and Medicine* in 2005 [60: 1107-1116].

throughs in the alcohol field typically reveal an inability to achieve sufficient replication (Peele, 1990); yet, these follow-up stories are less likely to be found on the front page and are usually relegated to the latter pages of the newspaper (Conrad & Weinberg, 1996). The question then becomes, why are there fewer social science "break-throughs" and why do they not receive similar coverage in the media? Some have argued that biomedical breakthroughs are more important and carry with them the hope that biomedical solutions will soon follow that will decrease and possibly prevent problems associated with alcohol use. A 1996 Alcohol Alert from the US federal agency on alcohol problems, the National Institute on Alcohol Abuse and Alcoholism (NIAAA), puts the argument succinctly: "...research suggests that the processes leading to the development of alcoholism reside largely in the brain. This has led to the concept of developing medications that act on specific brain chemicals to interfere with these processes." It was thus logical for the then-Director of NIAAA, Enoch Gordis, to conclude in the same publication that "developing pharmacotherapies for alcoholism treatment is a top priority of alcohol research" (NIAAA, 1996).

Another argument can be made that social scientists are less accustomed to promoting their own work and thus are less likely to advocate for our achievements. In the alcohol field in particular, biomedical findings are also seen by many as a validation of a particular governing image of alcohol problems, including alcoholism as a disease. Biological researchers commonly play into this hand: "I'm fully expecting ... to redefine alcoholism based on the underlying biological process and what it is about alcoholism that's inherited," one researcher told a reporter about his work (Harper, 2001).

One explanation for this lack of recognition is that many social science achievements are not made explicit in a format that is accessible to the larger public. More often results of social science research in the alcohol field are confined to specialized, professional alcohol, tobacco, and other drugs journals. While these journals are accessible to social science professionals working in the alcohol field, other disciplines and the general public are not the typical readers. Moreover, breakthroughs, as discussed in the media, are more likely to come from medical journals such as the Journal of the American Medical Association, the British Medical Journal, Lancet, and the New England Journal of Medicine. Because the orientation of these journals is predominantly medical, social science research is less likely to be published in these venues.

In this paper is we discuss four specific social science contributions to the alcohol field that have played a critical role in shaping its direction in research and policy. These contributions are in the areas of alcohol epidemiology, alcohol's contribution to the burden of disease, alcohol control policy, and screening/brief interventions. These four areas are not mutually exclusive, and neither do they exhaust the list of such contributions, but are offered as representative examples.

Alcohol Epidemiology: The Importance of Drinking Patterns

Epidemiology, the study of diseases and related conditions in populations, is a core concern of public health studies to which both medical and social science researchers regularly contribute. A distinction has commonly been drawn in alcohol studies between medical and social epidemiology (Caetano, 1991; Grant, 1994); here our concern is with the social epidemiological thread. Attempts to describe individual level drinking in populations date back in the U.S. to the Gallup Poll of 1939 when, shortly after the end of Prohibition, respondents were asked "Do you have the occasion to use alcoholic beverages such as liquor, wine, or beer or are you a total abstainer?" (Gallup, 1972). At that time, 58 percent of the U.S. population were defined by their answer to this question as current drinkers. While other surveys did occur between 1939 and the 1960s in the U.S. (Riley & Marden, 1947), they generally continued to focus on establishing the proportion of drinkers versus abstainers, or at most the frequency of drinking. The one exception was Straus and Bacon's (1953) include two dimensions of alcohol consumption (frequency and quantity), both of which continue to be important components of measuring alcohol use today.

The first national survey of alcohol use in a representative sample of the U.S. was conducted in the mid-1960s (Cahalan, Cisin, & Crossley, 1969). Early alcohol survey work pointed to the importance of assessing not just frequency of use or volume of use (frequency multiplied by usual quantity) but occasions of heavy episodic drinking (Knupfer, 1966). Within the first national U.S. survey, Cahalan et al. (1969) included an index that differentiated different types of drinkers based on the whether respondents "spaced" their drinking (drank fairly equal amounts regularly e.g., 2 drinks daily) or "massed" their drinking (drank large quantities at one time, e.g., 10 drinks on one day). Using a typology that included average volume and two levels of variability, the authors concluded that "drinkers who mass their drinks...appear to have more alcohol-related problems than do drinkers of equivalent over-all volume who space their drinks out over more occasions" (Cahalan et al. 1969, p 200). While many researchers in the social epidemiological tradition of alcohol studies continued to use typologies that distinguished between different patterns of drinking, there was a drift in the 1980s towards analyses of population surveys of drinking only in terms of the volume of drinking, that is, how much alcohol was consumed in some time frame. This was epitomized by the shift in meaning of "Q-F" (quantity-frequency) measures (Room, 1990): Q-F had originally referred to a typology of drinking patterns (Straus & Bacon, 1953), but came to refer simply to a volume measure. The drift in survey analyses was heavily influenced by the medical epidemiological literature, which invariably used a single continuous measure of volume, easily conducive to multivariate modeling of disease outcomes. Using a single continuous measure of alcohol use (volume) also fit well with other risk factors that are measured in specified units, e.g., tobacco. Furthermore, attention to the distribution of consumption or "total consumption" model (Ledermann, 1956), which was based on sales data without attention to patterning, also diverted research attention away from patterns during the 1970s and the 1980s.

By the 1990s, patterning of alcohol use was "rediscovered" and the use of measures to assess heavier episodic drinking began to be analyzed and reported in medical epidemiological studies as well as social research, since it has become apparent that pattern of drinking is important in the risk of chronic physical illnesses as well as in casualties and social problems. The reemphasis on patterns also reflected reactions by some researchers (Parker & Harman, 1978) and by alcoholic beverage industry advocates (Grant & Litvak, 1998) against the total consumption model. Over time, the typologies developed by Cahalan et al. (1969) were replaced by simpler, more direct measures of heavier drinking episodes (5 or more drinks on one occasion, during one day or at one time) that have now become standard in most studies that measure alcohol use (Grant, 2003; Wechsler, Dowdall, Davenport & Rimm, 1995).

Internationally, there are several comparative studies that have assessed drinking patterns across countries (Ager et al., 1996; Hupkens, Knibbe, & Drop, 1993; Leifman, 2002) and have specifically addressed demographic correlates of drinking patterns, primarily gender and age (Ager et al., 1996; Ahlström et al., 2001;

Johnstone, Leino, Ager, Ferrer, & Fillmore, 1996; Leifman, 2002). In addition, two international documents have recently been published that have provided guidelines for measuring and monitoring alcohol internationally and have emphasized the importance of assessing patterns of alcohol use (Dawson & Room, 2000; Stockwell & Chikritzhs, 2000).

Developments in comparative epidemiology in recent years have underlined the importance of customary patterns of drinking at the population level in understanding alcohol's role in harm to health (Room, 2001). For example, time-series analyses in western European countries of alcohol's role in total mortality suggest that an extra litre of pure alcohol per adult each year has three times the adverse effect on general mortality in northern Europe that it has in southern Europe (Norström, 2002). This north-south differential is found even for causes of death, such as cirrhosis (Ramstedt, 2001), where pattern of drinking had been thought to be irrelevant. Such findings have focused new attention on differences in patterns of drinking (Leifman, 2002), as the leading candidate for explaining these differences.

It is the social research traditions in alcohol studies, then, which first focused on patterns of drinking, and which have provided the tools for measuring and understanding the joint role of patterns, along with the level of consumption, in the occurrence of alcohol problems. Along with the focus on patterning of drinking has gone a new emphasis on the context of drinking, and particularly on the context of heavy drinking occasions, as a focus for policy efforts to reduce drinking-related harms (Babor et al., 2003).

Alcohol's Contribution to the Burden of Disease

An area that is developing rapidly in relatively recent times in which social scientists have played a major role is the assessment of the contribution of alcohol to morbidity and mortality from specific diseases and conditions within countries and globally. While technically this area is an extension of alcohol epidemiology, its growing importance deserves to be highlighted. Cross-sectional and longitudinal studies of alcohol's role in the development of and death from a range of diseases and conditions have been conducted for many years (Greenfield, 2001). There was also an early focus on determining the economic costs of alcohol abuse and alcoholism in the U.S. by assessing the costs attributable to alcohol in medical and health services, and from motor vehicle crashes, fire, crime and the social responses to alcohol abuse (Berry & Boland, 1977). This and later studies in the "costs of illness" tradition have been conducted by economists, but have depended heavily on an underlay of epidemiological studies. It was clear already at the time of Berry and Boland's study that a very large part of alcohol's health and social impact was in the form of casualties—injury and deaths both from accidents and from deliberate acts. An early effort to summarize this literature, conducted by a social science team (Aarens et al., 1977), resisted the tendency to develop one specific attributable fraction due to alcohol for each disease or condition, because they recognized dramatic variations in findings from different studies on the role of alcohol as a factor in morbidity and mortality, and that cultural differences in drinking patterns were one source of this variation. However, the "cost of illness" tradition and reviews in the medical epidemiological tradition continued to operate on the assumption that a single alcohol-attributable fraction (AAF) could be found for each cause of death or injury. For example Shultz, Rice et al. (1991) estimated single alcohol-attributable fractions, "the proportion by which disease cases, injury events, or deaths would be reduced if alcohol use and misuse were eliminated" (p 444), for each of a series of 35 diagnoses associated with alcohol use and misuse, based on previous studies (Rice, Kelman, & Miller, 1991). Their purpose was to develop software (ARDI—Alcohol-Related Disease Impact) so that individual U.S. states could separately calculate mortality, years of life lost, direct health care costs, indirect morbidity and mortality costs, and nonhealth-sector costs associated with alcohol use and misuse. While age and gender-specific rates were also calculated, AAFs, as derived from Shultz, Rice et al. (1991), were based on measures of volume of drinking only.

In 1995, another large medical epidemiological study was conducted, entitled "The Quantification of drug caused morbidity and mortality in Australia" (English et al., 1995), which also aimed at developing AAFs to estimate a range of outcomes including deaths, person-years of life lost, hospital episodes and bed days attributed to alcohol use. Whereas the Shultz et al. (1991) study relied on clinical case series studies, injury surveillance studies and specific epidemiological studies, in their report English et al. (1995) primarily used meta-analyses to determine these fractions for more than 100 diseases and conditions (both acute and chronic), assessing not only the role of alcohol but also of cigarette smoking and illicit drugs. Conditions were selected based on "any condition for which the literature search uncovered scientific evidence of an effect of alcohol, either harmful or protective, or at least the existence of a plausible hypothesis" (p. 65). Not surprisingly, given the way in which alcohol use is typically measured in the underlying studies, the authors used volume as the measure of alcohol consumption (categorized as hazardous and harmful versus low alcohol consumption), and omitted from their analyses the categories of ex-drinkers and binge drinkers when they were identified in specific studies. Patterns of alcohol use were thus not considered in their analyses.

However, there is increasing recognition that it is important to examine patterns of alcohol use as opposed to volume, given the strong relationship between heavier drinking patterns and morbidity and mortality (Rehm, Greenfield, & Rogers, 2001). In the light of this recognition, the most recent work at the World Health Organization (WHO) on estimating the alcohol's contribution to the global burden of disease (Ezzati et al., 2002) has taken account of variations in patterns between countries in estimating AAFs for all injuries (accidental and intentional) and for heart disease. Updating to 2000 earlier work on the global burden of disease for 1990 (Murray & Lopez, 1996a; 1996b), this study includes an assessment of the risk of alcohol for more than 60 ICD-10 codes for chronic diseases and injuries (Rehm et al., 2003). Most importantly, the recent analysis assessed alcohol use in two ways: average of consumption and patterns of drinking. Drinking patterns were estimated for each country by using expert informants and general population surveys (where available), in terms of proportions of abstainers, participation in heavy drinking occasions, drinking with meals and drinking in public places. Interestingly, the correlation between volume and drinking pattern was not significant across cultures, suggesting that drinking patterns measure a separate and unique dimension of risk.

Within the U.S., efforts are currently being made by the Centers for Disease Control and Prevention (CDC) to update the Alcohol-Related Disease Impact (ARDI) software program, developed over 10 years ago, which allows states as well as local health agencies to calculate mortality, years of life lost, and indirect mortality costs. ARDI has been updated with data from English et al. (1995) as well as incorporate findings from newer meta-analytic studies where available in order to estimate AAFs for both chronic and acute conditions related to alcohol use (Midanik et al., 2004). In 2002, NIAAA awarded a contract to the Harvard University School of Public Health to provide updated meta-analyses of epidemiologic studies of the association between alcohol consumption and a wide variety of health conditions and to estimate the fractions of U.S. morbidity and mortality from various conditions which are attributable to alcohol consumption.

The tradition of cost-of-illness estimates of alcohol's burden on society has been critiqued both in empirical and conceptual terms (Heien & Pittman, 1989), and there is no doubt that the new global-burden estimates will also be critiqued and improved over time. Nevertheless, there are three main ways in which the traditions are important in public health policymaking. First, as the traditions have developed, they both provide a basis for comparative evaluation of the relative role of alcohol and other risk factors. U.S. cost-of-illness studies, for instance, have consistently found that the economic costs of alcohol to be greater than the economic costs of illicit drugs (Harwood, Fountain, & Livermore, 1998), despite the much greater attention to drugs in political and public discourse. Likewise, the new WHO Global Burden of Disease (GBD) estimates have pointed new attention to the importance of alcohol as a risk factor in death and disability, with alcohol ranking number three among risk factors in developed societies, and number one among developing countries which are doing well (i.e., with low overall mortality—Ezzati et al., 2002). Second, both traditions, and particularly the burden-of-disease tradition, provide a basis for priority-setting in public health efforts to reduce harms from drinking. The GBD findings, for instance, reemphasize the importance of alcohol's role in casualties as a major component of years of life lost of impaired because of drinking. Third, making the estimates poses demands for data which have resulted in the development of new techniques and analyses, and which identify strategic gaps in the available information. Both traditions have thus helped to open up new research ideas and fields.

Estimating the burden of disease or social problems from alcohol is an example of a research enterprise that is necessarily multidisciplinary. It would make no sense at all for medical researchers or social researchers to ignore each other's work, and the estimation tasks in this field have in fact forced attention across literatures. The cost-of-illness studies have been led by economists, but

have necessarily depended heavily on the epidemiological literatures. While the overall impetus for the GBD approach owes more
probably to medical than to social traditions in public health, it is
notable that the working group on alcohol for the 2000 estimates
was primarily composed of social scientists (see list in Ezzati et al.,
2002).

Alcohol Control Policy

While there is an extensive history of attempts to control and prevent alcohol use and alcohol-related problems by governments, the
evaluation of alcohol policies and the use of alcohol research to
inform policy are fairly recent and date back to "the rise of modern
medicine and the emergence of the world Temperance Movement
during the 19th century" (Babor, 2002). The first international book
in the modern, post-temperance era to take on the challenge of finding ways to prevent alcohol problems on a societal or global level
was Alcohol Control Policies in Public Health Perspective (Bruun et
al., 1975), also known as the "Purple Book," published under the
auspices of WHO's European Office. This book represented a consensus of international experts on relevant approaches for the prevention of alcohol problems cross-nationally. It primarily focused
on the prevention of alcohol problems through reducing average
alcohol consumption by limiting access and availability. Rather than
focus on finding and providing treatment for heavier, problem drinkers as previous efforts had done, this group of scholars argued for
strategies that influence all drinkers, with the aim of reducing overall drinking. While the merits of this approach have been debated in
the literature over the years (Rehm, 1999; Room & Bullock, 2002;
Stockwell, Single, Hawks, & Rehm, 1997), there is little disagreement these days with the basic assertion of the model—based initially on a model often identified with Ledermann (1956), as developed in numerous studies conducted in Canada by Wolfgang
Schmidt, Reginald Smart and others—that the level of many alcohol
problems in a society, particularly cirrhosis deaths, drinking driving
casualties and violent crimes, tends to rise and fall with the overall
volume of consumption in that society (Cook, 1981; Cook & Moore,
1993). Moreover, it is clear that taxes and other alcohol control measures can have an influence on the overall level of consumption in a
population.

The analysis in the "purple book" drew on an already growing Nordic tradition of studies of the effects of taxes and other alcohol availability controls. In the 1970s, traditions of such studies began to get under way also in North America (Room, 1984; 2003), initially particularly focused on the effects of lowering and then raising the minimum drinking age in many Canadian provinces and U.S. states. Evaluation studies of other alcohol prevention and policy measures—e.g. of school-based education, of community prevention projects, and of drinking-driving countermeasures—began to appear and to proliferate. A U.S. National Academy of Sciences study (Moore & Gerstein, 1981) drew from the work of these nascent traditions together, and provided a stimulus for further work.

In 1994 international efforts were made to update the Purple Book and expand it, given new research that had appeared during the almost 20-year span. In Alcohol Policy and the Public Good (Edwards et al., 1994), attention was given on the one hand to epidemiological findings on both aggregate and individual risks of alcohol problems, and on the other hand to a somewhat broader range of prevention and policy strategies than in the "purple book"—including not only the influence of price and taxation of alcoholic beverages as prevention strategies and restrictions on access and availability of alcohol as a form of alcohol control, but also the measures affecting public safety and the social contexts in which problematic alcohol use occurs, information strategies, e.g., advertising, and their effects on alcohol use, and the use of interventions and treatment as a potential means of reducing rates of alcohol-related problems in the population. Other work in this area has appeared that focuses on alcohol control measures from the perspective of ways to minimize the harm of alcohol use (Plant, Single & Stockwell, 1997). These strategies include more global measures such as taxation policies (Godfrey, 1997) and standard unit labeling (Stockwell & Single, 1997) as well as more local level policies such as creating safer bars (Graham & Homel, 1997) and licensing regulations (Rydon & Stockwell, 1997).

More recently a group of 12 international alcohol scholars have assessed alcohol policy from the perspective of developing countries (Room et al., 2002). After evaluating the extent to which morbidity and mortality within developing countries is attributable to alcohol use, this group drew on the literature primarily from developed societies, and on a series of case reports, to outline specific approaches that can be used by governments to lessen the burden of alcohol. These approaches ranged from regulating access and availability of alcohol to harm reduction approaches such as helmet laws for motorcycles.

A third volume in the tradition of the Purple Book appeared in 2003, including a further broadening of the scope of coverage, in terms of different strategies of prevention. For the first time, the book includes a comparative evaluation of different prevention strategies in terms of their degree of effectiveness, the strength and cross-cultural breadth of the literature evaluating each strategy, and how inexpensive each is to implement (Babor et al., 2003).

During the past 30 years, we have experienced the growth of a substantial literature on the effects of different policies and interventions to reduce rates of alcohol-related problems. While a number of different disciplines have contributed to this literature, there is no doubt that the primary contributions have come from social scientists. In terms of pointing the way forward on effective polices and prevention strategies, this literature has put the alcohol problems field in a comparatively advantageous position compared with many other public health and social problems. Recent collaborative studies have brought political scientists' participation into the alcohol policy studies field (Sulkunen, Sutton, Tigerstedt, & Warpenius, 2000; Giesbrecht, Demers, Ogborne, Room, & Stoduto, in press; Greenfield et al., 1999). However, as was noted already at the time of the "purple book", there remains a substantial gap between our knowledge of effective strategies and the political will to implement them. Much of our knowledge on the effectiveness of particular strategies, indeed, has come from studies of what happened when they were dismantled (Olsson, Olafsdottir, & Room, 2002). However, as governments become increasingly concerned with revenue yield, strategies that involve increasing alcohol taxes may become more politically appealing assuming that alcohol beverages do not become too price responsive (Godfrey 1997).

Brief Interventions

Forty years ago, the primary treatment modalities for alcoholism (which included all kinds of alcohol problems) mostly involved relatively lengthy inpatient treatment. In the U.S., these included treatment in a mental hospital, in halfway houses, or in correctional facilities such as jail farms. Those receiving treatment were quite an extreme population, not only in terms of how heavily they had been drinking for how long, but also in terms of being without family and without work.

There have been enormous changes in treatment modalities and in the treatment system for alcohol problems since those days. On the one hand, there has been an enormous growth in the treatment system. On the other hand, the length of an episode of inpatient treatment has fallen, and treatment and intervention have increasingly moved to an outpatient basis. With the move towards screening and brief intervention in emergency rooms and other venues of general health and social service provision, the diffusion and transformation of alcohol treatment has now entered a new phase.

Social scientists have been deeply involved in developing an evidence base for alcohol treatment. The first national evaluation in the U.S. of the effectiveness of alcohol treatment, often known as the RAND study, was carried out by three social scientists (Armor, Polich, & Stambul, 1978; Polich, Armor, & Braiker, 1981). Social scientists have led the way in developing comparative evaluations of the effectiveness of different treatments (Finney & Monahan, 1996; Holder, Longabaugh, Miller, & Rubonis, 1991), including the landmark Project MATCH study (Project MATCH Research Group, 1997, 1998).

A major part of the change in alcohol treatment has been the rise of brief-intervention therapy—a shorter, time-limited strategy which can be used to lower alcohol consumption in at-risk populations and may also be useful to motivate towards treatment those individuals who are alcohol dependent. Within the context of healthcare delivery systems, brief interventions typically follow three steps: (1) A concise statement of the provider's concerns about the patient's current drinking and its affects on the patient's health; (2) Advice given to the patient which may be abstention or cutting down on alcohol use; and, (3) A conjoint decision on a strategy or a plan in order to achieve this goal (NIAAA, 1995). Probably one of the earliest examples of brief intervention, termed "advice," was a study conducted in 1977 by Griffith Edwards, Jim Orford and colleagues entitled "Alcoholism: A Controlled Trial of 'Treatment' and 'Advice.'" With an clinical sample, Edwards, Orford et al. (1977) provided strong evidence that brief treatment yielded similar outcomes as longer-term inpatient and outpatient care. Using a clinical trial of 100 alcoholic men who were married or living with someone, no significant differences were found for those respondents who received only one counseling session following assessment (advice) as compared to respondents who received more extensive treatment over the last year. The outcome measures (last 12 months) included drinking behavior, drinking problems, social adjustment and treatment experience.

The results of this research sent a very powerful message that is still felt today. Edwards et al. (1977) "set the stage" by demonstrating that no one particular treatment was superior to another. These findings continue to be replicated in larger studies, most recently in a clinical trial within the U.S. (Project MATCH Research Group, 1997, 1998). Most importantly, Edwards et al. (1977) also found that the duration of treatment was not related to outcomes even after one year. Finally, this research also demonstrated that "readiness" for treatment or what Edwards et al. (1977) termed "help-seeking behavior" is a very powerful determinant for how well a client does in treatment.

In the years that followed the Edwards et al. (1977) study, evaluation of brief interventions has become an arena in which social scientists have had a major role. Brief interventions have been examined in a variety of settings including hospitals, primary care clinics, college campuses, clinical research settings, and urgent care settings (U.S. Depart. of Health and Human Services, 2000). Several reviews of this literature have been written (Babor, 1994; Bien, Miller, & Tonigan, 1993; Chick, 1993) including one that focuses primarily on providing reading materials only (bibliotherapy) (Apodaca & Miller, 2003). In general, these reviews have found that brief interventions yield favorable outcomes when compared with control groups and longer counseling sessions.

Motivational interviewing, a type of brief intervention, is an area within the alcohol field that has had increased attention in the literature. Essentially, it is defined as a short-term assessment that involves personalized feedback on use of alcohol, its possible effects on physical and emotional health, and typically a discussion of possible ways to change their drinking behavior (Miller & Rollnick, 1991). This type of intervention has been used in a wide range of populations including psychiatric in-patients (Baker et al., 2002), college students (Borsari & Carey, 2000), veterans (Davis, Baer, Saxon, & Kivlahan, 2003), and older adults (Fleming, 2002).

Concerning developments in alcohol treatment, the attention of the mass media tends to be fixated on new medications or the prospect of gene manipulations, although the record suggests that in the alcohol field, when useful medications are found, they tend not to be used for ideological and other reasons (Room, 2004). It can be argued that a much more important development in alcohol treatment in the last 30 years has been the quiet evolution towards evidence-based therapies, notably including brief cognitive behavioral interventions, with social scientists taking the lead.

Beyond Dichotomies

While the focus of this chapter thus far has been on social science achievements, we do not mean to argue that the contributions of biomedical research are not important in the alcohol field in the understanding of the causes of alcohol dependence, its identification, and to some extent, its treatment. Rather, the aim of this book in general, and this chapter particularly has been is to work towards a balance of appreciation for individual biomedical and genetic factors that may be associated with alcohol dependence and the larger societal issues that influence drinking behavior. In order to achieve this balance, social science research in the alcohol field has been "showcased." The purpose was to accumulate in one place some of the major contributions of social science to the greater understanding of the impact of environmental factors in identifying, treating and perhaps more importantly, preventing alcohol-related problems.

However, there is a growing need to integrate the fields so that differences among fields can be appreciated instead of being viewed as a hindrance. An appreciation for this expansion was expressed almost a decade ago by the Gulbenkian Commission on the Restructuring of the Social Sciences (Wallerstein, 1996). This Commission, created in 1993 and composed of six scholars in the social sciences, two in the natural sciences, and two in the humanities, attempted in a series of three plenary meetings to consider what forces were influential in the current structure of the social sciences, and to recommend how they can be restructured and reorganized in the future. Based on their meetings and a report that was written in 1996 entitled *Open the Social Sciences*, the Commission made four recommendations: 1) "The expansion of institutions, within or allied to universities, which would bring together scholars for a year's work in common around specific urgent themes" (p 103); 2) "The establishment of integrated research programs within university structures that cut across traditional lines, have specific intellectual objectives,

and have funds for a limited period of time (say about five years)"
(p 103-104); 3) "The compulsory joint appointments of professors"
(p 104); 4) "Joint work for graduate students" (p 105).

Within each of these recommendations are examples of how broadening the research and teaching mission of the university across disciplines can only enhance progress from multiple levels. The Commission provides an example under their first recommendation of how a program created in Germany has brought together scholars on topics such as sociological and biological models of change.

Conclusions

There have been several challenges in writing this chapter. First, because its main purpose was to provide a broader view of the accomplishments of social scientists in the alcohol field, it has been difficult to maintain a balance between including too little or too much detail as each of these areas are described. In addition, it was difficult to decide which contributions should or should not be included. We recognize that by confining our discussion to four main areas, we have chosen not to focus on other meaningful contributions that deserve recognition. Among these are social scientists' contributions to the understanding of Alcoholics Anonymous (AA) which include international efforts to describe the processes by which AA meetings operate (e.g., Mäkelä, 1991;1994). In addition, social scientists have made significant contributions in general to the ways in which alcohol problems are conceptualized defined and discussed and specifically to the application of sociological constructivisim to the alcohol field (Gusfield, 1996; Room, 2001a). Finally, it is important to recognize the role of social science in the development and evaluation of employee assistance programs which initially focused on substance abuse issues and have later expanded to broader workplace issues (Roman, 1988; Roman & Blum, 2002).

Beyond this chapter, it has indeed been a major challenge to write a book that attempts to understand and translate larger trends in the health arena that directly affect the alcohol field. While I am left with many more questions than answers, nonetheless specific recommendations can be made. First, interdisciplinary research should be encouraged, fostered, and funded within universities and other research institutions and from public and private funders. Serious efforts on the part of mainstream institutions and funding agencies can create the conditions necessary to begin the process of unifying research efforts by encouraging research that crosses traditional disciplinary boundaries. Moreover, efforts can be made to find publication venues that encourage these interdisciplinary efforts. Currently, addiction journals published more psychosocial articles while disciplinary journals publish a greater proportion of biomedical articles in the alcohol field (Babor, Morisano, & Steinius, 2004). Second, similar to recent policies enacted by NIH to alleviate conflicts of interest of staff who serve as consultants or otherwise have outside interests in pharmaceutical and biotechology companies (Weiss, 2004), more recognition and a better understanding is needed of the role and the influence of vested interest groups such as the alcohol beverage industry on prevention and research efforts (Giesbrecht, 2000). Historically, the alcohol beverage industry has maintained powerful lobbies in Washington, D.C. that have influenced alcohol policy on the federal level (Greenfield et al., 1999). This power is felt most strongly whenever NIAAA recommends policies that may lower alcohol use in the population or encourage less advertising or target marketing of alcoholic beverages (Cahalan, 1987). Finally, it is critical that results of studies from a wide range of disciplines be given attention in the media so that breakthroughs are not solely defined by biomedical studies. This would involve educating journalists, for example, on stories that reflect the wide range of alcohol issues and policies that include multiple disciplines. One such effort is the Addiction Studies Program for Journalists of Wake Forest University that "…conducts workshops to explore the latest scientific research about addiction to cocaine, alcohol, nicotine, and other drugs. Workshops are held for reporters who cover all beats—science, medicine, and health, as well as crime and courts, sports, entertainment, education, and business" (http://www.addictionstudies.org). While this particular program tends to focus more on drugs and brain issues, it could potentially be expanded to include more social science materials.

Beyond the three recommendations above, it is essential that we continue to raise important questions that address the future of research in the alcohol field. Thus, I will delineate these areas of inquiry that I hope future scholars will embrace. First, to what extent can biomedicalization include multiple ways to address social and health problems? Is this process particularistic, in that it emerges in countries where social medicine has less of an impact or is it more universal, affecting other countries in perhaps different ways? Are there fields that are less impacted by the biomedical and biotechnical model, and if so, what makes these fields less susceptible to these influences? Finally, how much is advancement or progress in a field defined by the biomedical terminology and how much of this is the result of where agencies responsible for the problem are located and/ or the type of population involved? For example, it has recently been argued that in the last 15 years, biomedicalization in the field of aging has evolved into a new ethical arena in which life-extending interventions are increasingly becoming routine with less choice concerning whether or not a consumer wants to utilize these interventions (Kaufman, Shim, & Russ, 2004). It is my sincere hope that this book will help stimulate scholars from a wide range of disciplines to address these questions, and others as they arise, not only examine the effects of biomedicalization but also to work together to examine social issues from multiple viewpoints.

Note

1.   Portions of this chapter are reprinted with permission from Midanik, L.T. & Room, R. (2005). Contributions of social science to the alcohol field in an era of biomedicalization. Social Science and Medicine Social Science and Medicine, 60: 1107-1116.

References

Aarens, M., Blau, A., Buckley, S., Cameron, T., Goode, A., Lasser, E., et al. (1977). The epidemiological literature on alcohol, casualties and crime: systematic quantitative summaries. Berkeley: Social Research Group. 60, 1107-1116.

Ager, C. R., Ferrer, H. P., Fillmore, K. M., Golding, J. M., Leino, E. V., & Motoyoshi, M. (1996). Aggregate-level predictors of the prevalence of selected drinking patterns in multiple studies: A research synthesis from the Collaborative Alcohol-Related Longitudinal Project. Substance Use and Misuse, 31, 1503-1523.

Ahlström, S., Bloomfield, K., Knibbe, R., Allamani, A., Choquet, M., Cipriani, F., et al. (2001). Gender differences in the drinking patterns in nine European countries: descriptive findings. Substance Abuse, 22(1), 69-85.

Apodaca, T. R., & Miller, W. R. (2003). A meta-analysis of the effectiveness of bibliotherapy for alcohol problems. J Clin Psychol, 59(3), 289-304.

Armor, D. J., Polich, J. M., & Stambul, H. B. (1978). Alcoholism and treatment. New York: John Wiley and Sons.

Babor, T. F. (1994). Avoiding the horrid and beastly sin of drunkenness: Does dissuasion make a difference? Journal of Consulting and Clinical Psychology, 62(6), 1127-1140.

Babor, T. F. (2002). Linking science to policy: the role of international collaborative research. Alcohol Res Health, 26(1), 66-74.

Babor, T. F., Caetano, R., Casswell, S., Edwards, G., Giesbrecht, N. A., Graham, K., et al. (2003). Alcohol: No Ordinary Comodity. Research and public policy. New York, NY: Oxford University Press.

Babor, T. F., Morisano, D., & Steinius, K. (2004). How to choose a journal: Scientific and practical considerations. In T. F. Babor, K. Steinius & S. Savva (Eds.), Publishing Addiction Science: A Guide for the Perplexed (pp. 15-32). London, UK: International Society of Addiction Journal Editors.

Baker, A., Lewin, T., Reichler, H., Clancy, R., Carr, V., Garrett, R., et al. (2002). Evaluation of a motivational interview for substance use within psychiatric in-patient services. Addiction, 97(10), 1329-1337.

Berry, R. E., & Boland, J. P. (1977). Economic Cost of Alcohol Abuse. New York: Collier MacMillan.

Bien, T. H., Miller, W. R., & Tonigan, J. S. (1993). Brief interventions for alcohol problems: a review. Addiction, 88, 315-336.

Borsari, B., & Carey, K. B. (2000). Effects of a brief motivational intervention with college student drinkers. Journal of Consulting & Clinical Psychology, 68(4), 728-733.

Bruun, K., Edwards, G., Lumio, M., Mäkelä, K., Pan, L., Popham, R. E., et al. (1975). Alcohol Control Policies in Public Health Perspective (Vol. 25). Helsinki, Finland: The Finnish Foundation for Alcohol Studies.

Caetano, R. (1991). Psychiatric epidemiology and survey research: Contrasting approaches to the study of alcohol problems. Contemporary Drug Problems, 18, 99-120.

Cahalan, D. (1987). Understanding America's drinking problem: how to combat the hazards of alcohol. San Francisco: Jossey-Bass.

Cahalan, D., Cisin, I. H., & Crossley, H. M. (1969). American drinking practices: A national study of drinking behavior and attitudes (Vol. Monograph No. 6). New Brunswick, NJ: Rutgers Center of Alcohol Studies.

Chick, J. (1993). Brief interventions for alcohol misuse. British Medical Journal, 307(27), 1374.

Conrad, P., & Weinberg, D. (1996). Has the gene for alcoholism been discovered three times since 1980? A news media analysis. Perspectives on Social Problems, 8, 3-25.

Cook, P.J. (1981). The effect of liquor taxes on drinking, cirrhosis and auto fatalities. In M. Moore & D. Gerstein (Eds.). Alcohol and public policy: beyond the shadow of prohibition (pp. 255-285). Washington, DC: National Academy of Sciences.

Cook, P.J. & Moore, M.J. (1993). Violence reduction through restrictions on alcohol availability. Alcohol Health and Research World, 17, 151-156.

Davis, T. M., Baer, J. S., Saxon, A. J., & Kivlahan, D. R. (2003). Brief motivational feedback improves post-incarceration treatment contact among veterans with substance use disorders. Drug Alcohol Depend, 69(2), 197-203.

Dawson, D. A., & Room, R. (2000). Toward agreement on ways to measure and report drinking patterns and alcohol-related problems in adult general population surveys: The Skarpö Conference overview. Journal of Substance Abuse, 12, 1-21.

Edwards, G., Anderson, P., Babor, T. F., Casswell, S., Ferrence, R., Giesbrecht, N., et al. (1994). Alcohol Policy and the Public Good. Oxford, UK: Oxford University Press.

Edwards, G., Orford, J., Egert, S., Guthrie, A., Hawker, C., Hensman, M., et al. (1977). Alcoholism: a controlled trial of treatment and advice. Journal of Studies on Alcohol, 38, 1004-1031.

English, D. R., Holman, C. D. J., Milne, E., Winter, M. J., Hulse, G. K., Codde, G., et al. (1995). The Quantification of Drug Caused Morbidity and Mortality in Australia 1995. Canberra, Australia: Commonwealth Department of Human Services and Health.

Ezzati, M., Lopez, A. D., Rodgers, A., Vander Hoorn, S., Murray, C. J. L., & and the Comparative Risk Assessment Collaborating Group. (2002). Selected major risk factors and global and regional burden of disease. Lancet, 360, 1347-1360.

Finney, J. W., & Monahan, S. E. (1996). The cost-effectiveness of treatment for alcoholism: A second approximation. Journal of Studies on Alcohol, 57, 220-243.

Fleming, M. F. (2002). Identification and treatment of alcohol use disorders in older adults. In R. Atkinson (Ed.), Treating alcohol and drug abuse in the elderly (pp. 85-108). New York, NY: Springer Publishing Co, [URL:http://www.springerpub.com].

Gallup, G. (1972). The Gallup Poll: public opinion 1935-1971. New York: Random House.

Giesbrecht, N. (2000). Roles of commercial interests in alcohol policies: Recent developments in North America. Addiction, 95(4(Supp.)), 581-596.

Giesbrecht, N., Demers, A., Ogborne, A., Room, R., & Stoduto, G. (Eds.). (in press). Alcohol, Commerce and PublicHealth: Agendas in Recent Canadian Policy Experiences. Kingston: McGill-Queen's University Press.

Godfrey, C. (1997). Can tax be used to minimize harm? A health economist's perspective. In M. Plant, E. Single, & T. Stockwell (Eds.) Alcohol. Minimising the harm. (pp. 29-42). New York: Free Association Books.

Graham, K. & Homel, R. (1997) Creating safer bars. In M. Plant, E. Single, & T. Stockwell (Eds.) Alcohol Minimising the harm. (pp. 171-192). New York: Free Association Books.

Grant, B. F. (1994). Epidemiology. In J. W. Langenbucher, B. S. McCrady, W. Frankenstein & P. E. Nathan (Eds.), Annual review of addictions and treatment: volume 3 (pp. 71-86). New York: Pergamon Press.

Grant, B. F. (2003). Source and Accuracy Statement for the National Epidemiologic Survey on Alcohol and Related Conditions (NESARC), Wave I. Bethesda MD: National Institute on Alcohol Abuse and Alcoholism.

Grant, M., & Litvak, J. (Eds.). (1998). Drinking Patterns and Their Consequences. Washington, DC: Taylor and Francis.

Greenfield, T. K. (2001). Individual risk of alcohol-related disease and problems. In N. Heather, T. J. Peters & T. Stockwell (Eds.), International Handbook of Alcohol Problems and Dependence (pp. 413-437). New York: John Wiley.

Greenfield, T. K., Giesbrecht, N. A., Johnson, S. P., Kaskutas, L. A., Anglin, L. T., Kavanagh, L., et al. (1999). U.S. Federal Alcohol Control Policy Development: A manual. Berkeley, CA: Alcohol Research Group.

Gusfield, J. (1996). Contested meanings: The construction of alcohol problems. Madison: University of Wisconsin Press.

Harper, W. (2001, August 1). Strange brew. East Bay Express.

Harwood, H. J., Fountain, D., & Livermore, G. (1998). Economic costs of alcohol abuse and alcoholism. Recent Developments in Alcoholism, 14, 307-330.

Heien, D. M., & Pittman, D. J. (1989). Economic costs of alcohol abuse and drug abuse in the United States, 1992 [with commentaries]. Journal of Studies on Alcohol, 50(567-579).

Holder, H., Longabaugh, R., Miller, W. R., & Rubonis, A. V. (1991). The cost effectiveness of treatment for alcoholism: A first approximation. Journal of Studies on Alcohol, 52(6), 517-540.

Hupkens, C. L. H., Knibbe, R. A., & Drop, M. J. (1993). Alcohol consumption in the European community: Uniformity and diversity in drinking patterns. Addiction, 88, 1391-1404.

Johnstone, B. M., Leino, E. V., Ager, C. R., Ferrer, H., & Fillmore, K. M. (1996). Determinants of life-course variation in the frequency of alcohol consumption: Meta-analysis of studies from the collaborative alcohol-related longitudinal project. Journal of Studies on Alcohol, 57(5), 494-506.

Kaufman, S. R., Shim, J. K., & Russ, A. J. (2004). Revisiting the biomedicalization of aging: Clinical trends and ethical challenges. The Gerontologist, 44, 731-738.

Knupfer, G. (1966). Some methodological problems in the epidemiology of alcoholic beverage usage: the definition of amount of intake. American Journal of Public Health, 56(2), 237-242.

Ledermann, S. (1956). Alcool, alcoolisme, alcoolisation (Vol. 1). Paris: Presses Univarsitaires de France.

Leifman, H. (2002). Comparative analysis of drinking patterns in six EU countries in the year of 2000. Contemporary Drug Problems, 29(3), 501-548.

Mäkelä, K. (1991). Social and cultural preconditions of Alcoholics Anonymous (AA) and factors associated with the strength of AA. *British Journal of Addiction, 86*(11), 1405-1413.

Mäkelä, K. (1994). Rates of attrition among the membership of Alcoholics Anonymous in Finland. *Journal of Studies on Alcohol, 55*(1), 91-95.

Midanik, L. T., Chaloupka, F. J., Saitz, R., Toomey, T. L., Fellows, J. L., Dufour, M., et al. (2004). Alcohol-attributable deaths and years of potential life lost due to excessive alcohol use, United States, 2001. MMWR, 53(37), 866-870.

Miller, W. R., & Rollnick, S. (1991). Motivational Interviewing: Preparing People to Change Addictive Behaviour. New York: Guilford Press.

Moore, M., & Gerstein, D. (Eds.). (1981). Alcohol and public policy: beyond the shadow of prohibition. Washington, DC: National Academy Press.

Murray, C. J. L., & Lopez, A. (1996a). The Global Burden of Disease: A comprehensive assessment of mortality and disability from diseases, injuries and risk factors in 1990 and projected to 2020. Boston, MA: (Harvard School of Public Health on behalf of the World Health Organization and the World Bank).

Murray, C. J. L., & Lopez, A. D. (1996b). Quantifying the burden of disease and injury attributable to ten major risk factors. In C. J. L. Murray & A. D. Lopez (Eds.), The global burden of disease : a comprehensive assessment of mortality and disability from diseases, injuries, and risk factors in 1990 and projected to 2020 (Vol. 1, pp. 295-324). Cambridge, MA: Published by the Harvard School of Public Health on behalf of the World Health Organization and the World Bank.

NIAAA. (1995). The Physicians' Guide to Helping Patients with Alcohol Problems (No. NIH Publ. No 95-3769). Bethesda, MD: National Institute on Alcohol Abuse and Alcoholism.

NIAAA (1996). Alcohol Alert: Neuroscience research and Medications Development. No 33, July.

Norström, T. (Ed.). (2002). Alcohol in Postwar Europe: Consumption, Drinking Patterns, Consequences and Policy Response in 15 European Countries. Stockholm: National Institute of Public Health.

Olsson, B., Olafsdottir, H., & Room, R. (2002). Introduction: Nordic traditions of studying the impact of alcohol policies. In R. Room (Ed.), The Effects of Nordic Alchol Policies: What Happens to Drinking when Alcohol Controls Change? (Vol. NAD Publication 42). Helsinki: Nordic Council for Alcohol and Drug Research.

Parker, D. A., & Harman, M. S. (1978). The distribution of consumption model of prevention of alcohol problems: A critical assessment. Journal of Studies on Alcohol, 39, 377-399.

Peele, S. (1990). Second thoughts about a gene for alcoholism. The Atlantic Monthly, August, 52-58.

Plant, M., Single, E. & Stockwell, T. (Eds.) (1997) *Alcohol. Minimising the harm.* New York: Free Association Books.

Polich, J. M., Armor, D. J., & Braiker, H. B. (1981). The course of alcoholism: four years after treatment. New York: John Wiley and Sons.

Project MATCH Research Group. (1997). Matching alcoholism treatment to client heterogeneity: Project MATCH posttreatment drinking outcomes. Journal of Studies on Alcohol, 58(1), 7-29.

Project MATCH Research Group. (1998). Matching alcoholism treatments to client heterogeneity: Project MATCH three-year drinking outcomes. Alcoholism: Clinical and Experimental Research, 22(6), 1300-1311.

Ramstedt, M. (2001). Per capital alcohol consumption and liver cirrhosis mortality in 14 European countries. Addiction, 96(Suppl. 1), 19-34.

Rehm, J. (1999). Draining the ocean to prevent shark attacks? The empirical foundation of alcohol policy. Nordic Studies on Alcohol and Drugs, 16 (English Suppl.), 46-54.

Rehm, J., Greenfield, T. K., & Rogers, J. D. (2001). Average volume of alcohol consumption, patterns of drinking and all-cause mortality. Results from the U.S. National Alcohol Survey. American Journal of Epidemiology, 153(1), 64-71.

Rehm, J., Room, R., Monteiro, M., Gmel, G., Graham, K., Rehn, N., Sempos, C.T. & Jernigan, D. (2003). Alcohol as a risk factor for global burden of disease. *European Addiction Research, 9*(4), 157-164.

Rice, D. P., Kelman, S., & Miller, L. S. (1991). Estimates of economic costs of alcohol and drug abuse and mental illness, 1985 and 1988. Public Health Reports, 106(3), 280-291.

Riley, J., Jr., & Marden, C. (1947). The social pattern of alcoholic drinking. Quarterly Journal of Studies on Alcohol, 8, 265-273.

Roman, P. (1988). Growth and transformation in workplace alcoholism programming. *Recent Developments in Alcohol, 6,* 131-158.

Roman, P. & Blum, T. (2002). The workplace and alcohol problem prevention. *Alcohol Research and Health, 26*(1), 49-57.

Room, R. (1984). Alcohol control and public health. Annual Review of Public Health, 5, 293-317.

Room, R. (1990). Measuring alcohol consumption in the United States: methods and rationales. In L. Kozlowski, H. M. Annis, H. D. Cappell, F. B. Glaser, M. S. Goodstadt, Y. Israel, H. Kalant, E. M. Sellers & E. R. Vingilis (Eds.), Research advances in alcohol and drug problems (Vol. 10, pp. 39-80). New York: Plenum Press.

Room, R. (2001a). Governing images in public discourse about problematic drinking. In N. Healther, T.J. Peters & T. Stockwell (Eds.) *Handbook of alcohol dependence and alcohol-related problems* (pp. 33-45), Chichester UK: Wiley.

Room, R. (2001b). New findings in alcohol epidemiology. In N. Rehn, R. Room & G. Edwards (Eds.), Alcohol in the European Region: Consumption, harm and policies (pp. 35-42). Copenhagen: World Health Organization, Regional Office for Europe.

Room, R. (2003). Effects of alcohol controls: Lessons from a half-century of nordic social experimentation. Paper presented at the Preventing Substance Use, Risky Use, and Harm: What is Evidenced-Based Policy?, Fremantle, Western Australia.

Room, R. (2004). What if we found the magic bullet? Ideological and ethical constraints on biological alcohol research and its application. In H. Kleingemann (Ed.), From Science to Action? 100 Years Later - Alcohol Policies Revisited. (pp. 153-162). Dordrecht, The Netherlands: Kluwer.

Room, R., & Bullock, S. (2002). Can alcohol expectancies and attributions explain Western Europe's north–south gradient in alcohol's role in violence? Contemporary Drug Problems, 29, 619-648.

Room, R., Jernigan, D., Carlini-Marlatt, B., Gureje, O., Makela, K., Marshall, M., et al. (2002). Alcohol and the developing world: a public health perspective (Vol. 46): Finish Foundation for Alcohol Studies in collaboration with World Health Organization.

Rydon, P. & Stockwell, T. (1997). Local regulation and enforcement strategies for licensed premises. In M. Plant, E. Single & T. Stockwell (Eds.). *Alcohol. Minimising the harm* (pp. 211-229). New York: Free Association Books.

Shultz, J. M., Rice, D. P., Parker, D. L., Goodman, R. A., Stroh Jr., G., & Chalmers, N. (1991). Quantifying the disease impact of alcohol with ARDI software. Public Health Rep, 106(4), 443-450.

Stockwell, T., & Chikritzhs, T. (Eds.). (2000). International guide for monitoring alcohol consumption and related harm. WHO/MSD/00.4. Geneva: Department of Mental Health and Substance Dependence, Noncommunicable Diseases and Mental Health Cluster, World Health Organization.

Stockwell, T. & Single, E. (1997). Standard unit labeling of alcohol containers. In M. Plant, E. Single & T. Stockwell (Eds.). *Alcohol. Minimising the harm* (pp. 85-104). New York: Free Association Books.

Stockwell, T., Single, E., Hawks, D., & Rehm, J. (1997). Sharpening the focus of alcohol policy from aggregate consumption to harm and risk reduction. Addiction Research, 5(1), 1-9.

Straus, R., & Bacon, S. D. (1953). Drinking in college. New Haven: Yale University Press.

Sulkunen, P., Sutton, C., Tigerstedt, C., & Warpenius, K. (2000). Broken Spirits: Power and Ideas in Nordic Alcohol Control (Vol. Nad Publication No. 39). Helsinki: Nordic Council on Alcohol and Drug Research.

U.S. Depart. of Health and Human Services. (2000). Alcohol and Health. 10th Special Report to the U.S. Congress. Bethesda: MD: Public Health Services, National Institutes of Health, National Institute on Alcohol Abuse and Alcoholism.

Wallerstein, I. (1996). Open the Social Sciences. Report of the Gubenkian Commission on the Restructuring of the Social Sciences. Stanford: Stanford University Press.

Wechsler, H., Dowdall, G. W., Davenport, A., & Rimm, E. B. (1995). A gender-specific measure of binge drinking among college students. American Journal of Public Health, 85(7), 982-985.

Weiss, R. (2004, September 24, 2004). NIH bans collaboration with outside companies; Policy comes after conflict-of-interest inquiry. The Washington Post.

Subject Index

Name Index

167